Mechanics of Nanocomposites

Mechanics of Nanocomposites

Homogenization and Analysis

Authored by

Farzad Ebrahimi, Ali Dabbagh

CRC Press
Taylor & Francis Group
Boca Raton London New York

CRC Press is an imprint of the
Taylor & Francis Group, an **informa** business

Published 2020

by CRC Press
6000 Broken Sound Parkway NW, Suite 300, Boca Raton, FL 33487-2742
and by CRC Press
2 Park Square, Milton Park, Abingdon, Oxon, OX14 4RN

First issued in paperback 2021

© 2020 Taylor & Francis Group, LLC
CRC Press is an imprint of Taylor & Francis Group, an Informa business

No claim to original U.S. Government works

ISBN 13: 978-1-03-223575-2 (pbk)
ISBN 13: 978-0-367-25433-9 (hbk)

**Visit the Taylor & Francis Web site at
http://www.taylorandfrancis.com**

**and the CRC Press Web site at
http//www.crcpress.com**

Typeset in Times
by Cenveo® Publisher Services

Dedication

"To my honorable parents"

Farzad Ebrahimi

"To my dearests: my honorable parents and my beloved sisters"

Ali Dabbagh

Table of Contents

PART 3   *Dynamic Analyses*

Preface

Nowadays only a limited number of people are not yet familiar with the word *Nano*. This is due to the widespread application of the nanosize systems in different positions and devices which might be directly or indirectly involved with our lives. Nanosize devices reveal many enhanced mechanical, electrical, optical, and thermal behaviors that cannot be monitored at all once the bulk of the same material is implemented. Hence, nanosize elements can be desirable to the designers of modern complicated systems because of their remarkable potential, which encourage the engineers to implement such tiny devices in various industrial applications.

In modern engineering applications, often homogeneous materials cannot satisfy the purposes of a desired design. Therefore, engineers attempt to gain the capacities of various materials by manufacturing heterogeneous materials, named composites. As you know, one of the most crucial reasons of making composites is to fill in the weaknesses of a primary matrix with the strong corresponding features of the reinforcing phase. One of the most important issues to attract the attention of engineers is to produce stiffer materials using the concept of the composites. To this purpose, several fibers and particles have been used by many researchers. Utilization of nanosize particles (nanoparticles), nanosize fibers (nanofibers), and nanosize fillers (nanofillers) as the reinforcing phase dispersed in a primary matrix can improve the stiffness of the obtained composite more than when macro size reinforcing elements are used. This is why researchers brought the idea of fabricating nanocomposite (i.e., a composite with reinforcements having at least one dimension within the range of 1–50 nm) to life. The manufactured advanced materials are able to endure greater mechanical excitations due to their stiffer nature. For instance, nanocomposite structures can be better candidates to be employed in the applications of which the mechanical element is assumed to be subjected to critical buckling excitations.

In addition to the above positive aspects of nanocomposites, it must be pointed out that these materials can be involved with some destructive phenomena, such as existence of porosities due to errors in the fabrication process, which cannot be completely omitted. Also, polymeric nanocomposites will show a time-dependent constitutive behavior and it is important to consider the rheology of such materials. In the present book, the first chapter will be dedicated to the introduction and applications of nanocomposite materials. Afterward, the second chapter will present the mathematical tools required to carry out the analyses presented in the following parts of the book. In chapter 2, an introduction to the elasticity relations and different types of shell, beam, and plate theories will be presented followed by micromechanical homogenization of polymeric nanocomposites reinforced with carbon nanotubes, graphene oxides, and graphene platelets regarding the effects of the existence of porosities, the viscoelastic nature of the material, and the single- or multi-layered character of the nanocomposite structure. Following the first two chapters, chapters 3–10 examine the static analyses and chapters 11–21 discuss the dynamic responses of continuous systems manufactured from nanocomposites.

Farzad Ebrahimi
Imam Khomeini International University, Qazvin, Iran

Ali Dabbagh
University of Tehran, Tehran, Iran

Acknowledgment

The authors are eager to extend their gratitude to those who played significant roles in bringing this book to life. The authors would also like to thank the Taylor & Francis Group/CRC press for their invaluable support during the preparation of the book. The authors relied greatly upon their insight and kindness. The authors' special gratitude goes to Allison Shatkin and Gabrielle Vernachio, senior commissioning editor and editorial assistant of the CRC Press for their kind behavior during the long period of book preparation. Also, the authors are thankful for the efforts of Camilla Michael, the previous editorial assistant of the CRC Press for her valuable assistance. Furthermore, the first author would like to thank some of his students such as Saeideh Qaderi, Navid Farazmandnia, Mostafa Nouraei, Ali Seyfi, and Pendar Hafezi for their endeavors in the book's preparation procedure. The second author would like to thank the dear friends he has relied upon for their insight and expertise in his research; special thanks go to Sama Gharaei, Soroush Gharaei, and Amir Hojjatpour who selflessly contributed to this project through their linguistic and graphic knowledge.

About the Authors

Farzad Ebrahimi is an associate professor in the Department of Mechanical Engineering at the Imam Khomeini International University (IKIU), Qazvin, Iran. He received his Ph.D. from the School of Mechanical Engineering of the University of Tehran in 2011. Thereafter, he joined IKIU in 2012 as an assistant professor and became an associate professor in 2017. His research interests include the mechanics of nanostructures and nanocomposites, smart materials and structures, viscoelasticity, composite materials, functionally graded materials (FGMs), and continuum plate and shell theories. He has published more than 350 international research papers and is the author of three books about smart materials; he has also edited three books for international publishers. His recent book, titled *Wave Propagation Analysis of Smart Nanostructures*, was published in collaboration with Taylor & Francis Group/ CRC Press and is concerned with the investigation of the wave dispersion characteristics of smart nanostructures manufactured from composites.

Ali Dabbagh is studying for his M.Sc. in the School of Mechanical Engineering, College of Engineering, University of Tehran, Tehran, Iran. His research interests include solid mechanics, smart materials and structures, composites and nanocomposites, functionally graded materials (FGMs), nanostructures, and continuum plate and shell theories. He has published more than 15 international research papers in his research area and has authored one book. His book, titled *Wave Propagation Analysis of Smart Nanostructures*, deals with the wave dispersion characteristics of smart composite nanostructures and was published by Taylor & Francis Group/CRC Press. He is currently working on his M.Sc. thesis, concerned with the mechanical behaviors of hybrid nanocomposite structures subjected to various static and dynamic excitations.

Part 1

Preliminary Requirements

Part 1

Preliminary Requirements

1 Introduction

1.1 AN INTRODUCTION TO COMPOSITES AND NANOCOMPOSITES

Because it is very difficult to find a material with suitable stiffness, strength, low weight, and efficient thermal behaviors among the common homogeneous materials, engineers were motivated to design a new group of materials to be constructed from two or more phases stacked together with the goal of satisfying all of the design requirements. Bringing this idea to the life results in *composite* materials. Composite materials make it possible to tailor material properties, tuning the amount of each of the phases in the fabricated material. In general, three major kinds of composites can be formed. The first is *fibrous* composite, or *fiber-reinforced* composite (FRC), which can be achieved by scattering a group of fibers of material A in material B. The second type is the *particulate* composite, in which macroscale particles of material A are dispersed in material B. The last and third type belongs to a combination of the first two types. *Laminated* composites are constructed from a group of finite layers, either fibrous or particulate, stacked on top of each other. For example, a very famous and widespread example of fibrous composites is reinforced concrete. The major goal of fabrication of this composite was to create a material that enables a structure to endure tension as well as compression.

As stated in the final part of the previous paragraph, it is common to use a group of fibers dispersed in an initial material to improve the material properties of the host material to make an FRC. FRC is one of the most well-known types of composites and is broadly used in engineering designs. But, why use fibers in an initial matrix? The answer to this question may be better found looking at materials science references; however, a general answer to this question can be found by remembering that the material properties of small fibers are better than those of the same specimen in its bulk form (Reddy, 2004). Indeed, the fibers of a material can provide better stiffness and strength, as well as lower weight. Hence, fibers are widely implemented in FRCs to reinforce the host material. In the literature, the host material which is responsible for covering the fibers is called the *matrix* material. The main functionality of the matrix in the FRC is to protect the arrangement of the fibers in the composite and also to act as a load-transfer medium to avoid proximity of the fibers and the environment. It is worth mentioning that the implemented fibers are divided into two categories: short and long fibers. Short fibers, also called *whiskers*, are better alternatives for the purpose of reinforcing the matrix in various aspects. Fibers employed in the FRCs are generally slender fibers. In other words, the slenderness ratio (i.e., the length-to-diameter ratio) of these fibers is a very big number, for instance, from 10 to 100. Both metallic and non-metallic materials can be selected to play the role of either matrix or fiber in the FRC materials.

Furthermore, to enhance the entire toughness of the composite it is usual to use coupling doer in the laminated composites, consisting of FR laminae. In addition to adding the toughness of the composite, adding coupling doer will amplify the bonding between the fiber and matrix in each stacking ply of the composite. It is very important to disclose the method used to reinforce each of the stacking plies of a laminated composite consisting of FR plies, because the orientation of the fibers in the lamina can affect the mechanical behavior of the laminated composite. Each lamina can be reinforced by dispersing fibers in the matrix in four main types. The first type is *unidirectional* FRC which can be achieved by placing the fibers only in one direction in the matrix. This type of FRC can have remarkable stiffness, strength, toughness, thermal resistance, and so on only in the specified direction that the fibers were aligned. Dispersion of fibers in two desired directions in the matrix results in *bi-directional* FRCs, which have improved features in these directions. When fibers are scattered in the host matrix with arbitrary directions, *discontinuous* FRCs are obtained, whose mechanical properties cannot be estimated due to the random orientation angle of the fibers in the composite. However, it is obvious that the reinforcing procedure cannot be completed as well as continuous unidirectional and bi-directional composites. The last type of FRC is *woven* composite, which will be constructed implementing a group of wavy fibers reticulated in each other. Regarding the above types of composites, the functionality of the laminated composite can be influenced using each type of composite in the stacking plies. For example, the shear durability of the laminate can be improved by employing unidirectional FRCs with an orientation angle between $\pi/6$ and $\pi/4$ radians as the constituents. It is worth mentioning that in the case of using plies whose fibers orientation angles are perpendicular to each other, a cross-ply laminated composite is achieved. When the orientation angle of the fibers in each layer can be desirable, the obtained composite is an angle-ply one.

In the above paragraphs, various aspects of FRC and laminated composites were discussed, emphasizing the advantages of composites. However, implementing composites without paying attention to the practical problems in the fabrication procedures will result in unavoidable losses. First, consider that fiber debonding is one of the worst possibilities that can occur in composites. Also, delamination of the plies must be critically considered when designing a laminated composite for a desired application. Delamination of plies happens because of differences between the material properties of the layers. Indeed, this difference makes it possible for each layer to show a different behavior from itself subjected to shear stress in comparison with its neighbor layer(s). Therefore, the danger of delamination in the laminated composites must be included in the analysis of structures fabricated from such materials, especially at the edges. Another important issue is to consider the voids and porosities that may be generated in the composite during the fabrication process. Finally, it is noteworthy that sometimes the fibers utilized in the composite are internally damaged and this issue can dramatically affect the mechanical behavior of the composite.

It can be helpful to mention some of the common applications of composites in various engineering designs. Before beginning any discussion about practical applications of composites, it is essential to confess that due to the broad spectrum of applications of composites it is impossible to mention all their real-world applications in the present text. However, mentioning the most crucial and strategic applications of such complex materials will give a general knowledge about various fields in which composites are involved. Composites are widely implemented in the design of military equipment and aircraft. In military and aircraft applications, the low-weight-to-efficiency ratio is one of the most crucial requirements of a proper design. One of the primary applications of composites in this area is the employment of boron FRCs with an epoxy matrix in the horizontal stabilizers of the F-14 (Mallick, 2007). Later, carbon fibers were broadly implemented by engineers to design different components of aircrafts. For instance, the airframe and flap of the AV-8B, the outer skin of the B-2, the fin and under the wing fairings of the F-11, and the fin box and fin leading edge of the F-16 are manufactured from carbon fiber-reinforced composites (Mallick, 2007). In 1987, Airbus utilized composite materials as the constituent material of some parts of the A310 aircraft such as the lower access panel, top panels of the wing leading edge, outer deflector doors, engine cowling panels, elevators, and fin box (Mallick, 2007). After a year, Airbus presented its newly developed aircraft, the A320, with a fully composite tail. In 2006, Airbus released the A380 aircraft with 25% weight fabricated from composites. Some of the most important components of this aircraft made from composites are the tail, wing box, outer wing, flap track fairings, fixed leading edge, upper and lower panels, main landing gear leg fairing door, spoilers, pylon fairings, nacelles, and cowlings (Mallick, 2007). Application of composites in helicopters, as one of the most important military devices, can be observed in various parts of such machines like baggage doors, fairings, vertical fins, and tail rotor spars. Also, implementation of either carbon or glass FRCs in the rotor blades of helicopters can satisfy two major requirements: weight reduction, which is of great importance for such devices, and improved vibration control of the blades (Mallick, 2007). In fact, employment of FRCs enables the engineers to control dynamic fluctuations by changing the type, distribution, concentration, and fibers' orientation angle of the composite rather than the former, conventional use of control masses in blades made of aluminum alloys. In addition, using FRCs in the fabrication of the helicopters' components helps the manufacturer to generate more complicated structures due to the remarkable flexibility and forming potentials of composites in comparison with conventional metallic alloys (Mallick, 2007).

Composites have also been employed by many automotive companies for various components of their machines. Generally, composites are used in three major categories in the automotive industry: body, chassis, and engine. For example, E-glass fibers are utilized to construct sheet molding compound (SMC) composites to be used in the hood and door panels to enhance the stiffness in such components in addition to generate an increase in the damage tolerance of such parts (Mallick, 2007). The main reason for this choice, compared with other available composites like carbon fiber-reinforced composite materials, is the lower cost of this type of composite. It is noteworthy that SMCs are fabricated in a compression molding procedure that cannot satisfy the "Class A" surface finish required for exterior body components of machines. To solve this problem, it is common to use a flexible resin as the coating on the molded compound (Mallick, 2007). It is remarkable that using SMCs in various elements of mechanical machinery results in lower tooling costs and parts integration on top of the primary goals of using composites (i.e., weight reduction in the designed element). As a rough approximation, implementing SMCs instead of steel alloys can lessen the shaping costs by 40% to 60% (Mallick, 2007). A clear example of having suitable part integration in components manufactured from SMCs is the radiator of an automobile. In this example, the supports are made from SMCs that can be joined together easily using an adhesive instead of using a large number of screws. Chevrolet was the first automobile manufacturer to seriously use FRCs in the chassis during fabrication of the Corvette in 1981. In this attempt, the rear leaf springs were constructed from uni-leaf E-glass fiber-reinforced polymeric composites instead of multi-leaf ones. Due to this replacement, the entire weight of the rear leaf springs was decreased up to 20% (Beardmore, 1986). It must be noted that use of composites in the engine elements has not attracted designers' attention in the same way as designing body or chassis components. Indeed, the working condition of an engine requires a material that can endure dynamic fatigue loads in an extreme thermal environment. As well, implementation of composite materials for automotive industries with their voluminous production is different from their utilization in other industries like aircraft. This reality can be better perceived by remembering that the fabrication rate of aircraft during a year might be a few hundred specimens; however, this same number is manufactured within hours by automobile manufacturers (Mallick, 2007). In the automotive industries, composites' curing time is one of the most significant determining parameters in selecting suitable alternative for the designed structure. For example, epoxy matrix cannot suit automotive applications due to its time-consuming curing procedure, whereas polyester, vinyl ester, and polyurethane can be utilized as the matrix of the composite because their curing time is not as long as epoxy (Mallick, 2007). Another crucial criterion in collecting composites in the automotive industry is the suitable cost of the selected composite. For instance, it is not usual to use carbon fibers in the automotive industries because its manufacture is very expensive; however, E-glass fibers are often chosen due to their reasonable production costs. Still, although E-glass fiber is cheaper than carbon fiber, in some of the modern designs carbon fibers are used to produce a composite to be employed in the automotive industry. The reason is that the low-weight-to-strength and/or stiffness ratio that can be provided by carbon fiber-reinforced composites cannot be satisfied with glass fiber-reinforced ones (Mallick, 2007). For example, the roof panel of the BMW M6 is made from carbon fiber-reinforced composites.

In addition to the above applications in the military, aircraft, and automotive industries, another engineering field that requires composites is the field of aerospace. In aerospace applications, decreased weight combined with high mechanical, thermal, and electrical performance is an indispensable design requirement. Composites possess a wide range of applications in this area. In addition, it is common to observe huge thermal loadings in aerospace applications. Therefore, composites may be needed to improve the thermal resistance of the element to enable it to endure thermal loading. For example, implementing materials with a coefficient of thermal expansion (CTE) of nearly zero can enormously increase the stability of aerospace devices in a wide range of temperature increases. This purpose can be enhanced by tuning the volume fraction of the components involved in the fabrication of composites. Also, the superiority of composite materials to conventional homogeneous ones is that it is possible for composites to incorporate a remarkable mechanical stiffness and strength with suitable thermal resistance, for example by using carbon fiber-reinforced composites. In space telescopes, the mirrors and lenses are embedded into carbon fiber-reinforced composites (Wolff, 1979). Furthermore, the abovementioned type of composites can be utilized as the truss structures in low earth orbit (LEO) satellites and interplanetary satellites (Mallick, 2007). It must be noted that epoxy-based carbon fiber-reinforced composites are better candidates for space applications concerned with huge thermal loadings compared to metallic homogeneous materials or composites fabricated using a metal-based matrix. The reason for this is that using epoxy resin can guarantee enhancement of the mechanical performance as well as reaching zero CTE for the composite (Mallick, 2007).

It is worth mentioning that composites' applications are not limited to those reported in the above paragraphs, which were aimed at presenting general applications of composites in those engineering fields important for a mechanical engineer and all applications of composites were not included. In fact, the focus was on the major mechanics-based industries involved with composites. Readers are highly advised to look at other references in the composites' open literature to learn more about other applications of such crucial materials (Mallick, 2007). On the other hand, due to their wide application in different engineering fields such as mechanics, aerospace, civil, electronics, and so on, some researchers have made efforts to analyze the mechanical characteristics of composite materials and structures subjected to various types of environmental and/or external loadings. What follows is a general literature review on the activities performed by researchers in this area. Aref and Alampalli (2001) presented a finite element (FE) based numerical analysis dealing with the vibrational characteristics of a composite slab bridge fabricated in the United States and considered for various modes of the structures' fluctuation. They verified their numerical data with the experimental ones. The composite bridge has a longitudinal joint to act as the shear key of the bridge. The influence of the damages in the joint was included in their FE analysis. A general static and dynamic investigation was performed by Zenkour and Fares (2001) in the framework of a modified version of the first-order shear deformation theory (FSDT) of shells free from employment of shear correction coefficient to probe bending, buckling, and vibration responses of laminated composite shells. Eight-node serendipity elements were utilized by Patel et al. (2002) to capture the effects of placing the laminated composite plate in a hygrothermally influenced environment on the static and dynamic behaviors of the composite structure based on the fundamentals of the FE method (FEM) in association with a higher-order shear deformation theory (HSDT). The snap-through characteristics of laminated FRCs were explored by Dano and Hyer (2002) in the framework of the Rayleigh-Ritz method. They showed the validity of their investigation by comparing its results with experimental data. Furthermore, the compatibility of the composite structures with changes in working conditions was covered in an analysis carried out by Dano and Hyer (2003) using shape memory alloy (SMA) fibers in unsymmetric laminated composites to study the snap-through characteristics of such smart composites. In another paper, Wang et al. (2002) used a meshless kernel particle FE framework to solve the vibration problem of composite plates based upon the FSDT regarding various types of boundary conditions (BCs). Implementing various classical and shear deformable plate hypotheses for plate-type elements, Zenkour (2004) captured the influences of the time-dependency of the material properties of the composite structure on the quasi-static stability responses of the composite plates. The issue of solving the dynamic problem of moderately-thick composite plates was studied by Ferreira et al. (2005) based on the FSDT incorporated with the radial basis functions. Pelletier and Vel (2006) utilized the concept of the genetic algorithm (GA) for the goal of optimizing FR laminated composites to enhance a group of properties of the composite like stiffness, strength, and so on. Aydogdu (2007) utilized an FE-based Ritz-type solution for the goal of analyzing the thermo-elastic buckling problem of a cross-ply laminated composite beam considering the contact conditions between the stacking plies of the beam. The determination of the natural frequency and buckling load of cross-ply laminated composite shells was procured by Matsunaga (2007) regarding the effects of shear deformation on the mechanical behavior of the composite. Furthermore, both impact and vibrational behaviors of multi-layered honeycomb FR polymer composites were surveyed by Qiao and Yang (2007). The authors utilized commercial programs such as ABAQUS and LS-DYNA to verify their results with numerical simulations of FE-based programs. The application of a newly developed flexural modulus for investigating bending, buckling, and vibration problems of laminated composite beams was shown by Chai and Yap (2008). Balzani and Wagner (2008) probed the issue of appearance of delamination in unidirectional FR laminated composites. In another article, Aydogdu (2009) presented a novel shear deformable shape function for higher-order continuous systems to analyze the mechanical behaviors of laminated composite plates. He compared the results of his hypothesis with those enriched from the 3-dimensional (3-D) elasticity solution to select the best coefficient to use in his newly developed theory. The contact conditions between layers of composite plates were included by Ćetković and Vuksanović (2009) employing the concept of the well-known layer-wise theory (LWT) for the purpose of studying bending, buckling, and vibration problems of composite structures. Roy and Chakraborty (2009) used the fundamentals of the GA to control the dynamic response of FRC doubly-curved shells. The concept of Carrera's unified formulation (CUF) was employed by Fazzolari and Carrera (2011) to probe natural frequency and buckling load characteristics of composite plates. In another endeavor, CUF was implemented in association with the radial basis functions to procure a finite difference (FD)-based study dealing with the static and dynamic behaviors of laminated composite plates using a zig-zag plate model (Rodrigues et al., 2011). An FE-based isogeometric analysis (IGA) was conducted by Shojaee et al. (2012) dealing with both static and dynamic analyses of laminated composite plates. Thai et al. (2012) procured an IGA about the static and dynamic characteristics of laminated composite plates in the framework of the FSDT. Continuing with performing FE-based IGA about laminated composite plates, Thai et al. (2013) mixed the NURBS basic functions with the LWT for plates modeled by the FSDT. The nonlinear mechanical analysis of smart piezoelectric FRC beams was completed by Mareishi et al. (2014) based on the Euler-Bernoulli beam theory. Moreover, the vibrational characteristics of laminated composite plates were explored by Thai et al. (2014) based upon the well-known IGA in association with a newly introduced inverse tangential shape function to present a novel HSDT. The application of the third-order SDT (TSDT) of Reddy for solving bending, buckling, and free vibration problems of laminated composite plates was shown by Thai et al. (2015) within the framework of implementation of an IG-aided C^1-continuous procedure. On the other hand, Sepahvand (2016) considered the existence of uncertainties in the fibers' orientation angle while analyzing the natural frequency behaviors of FRC plates using spectral stochastic FEM. Yu et al. (2016) used a new, simple FSDT for composite plates with cutout to study the vibration and buckling problems of composite structures within the framework of an IGA. Alesadi et al. (2017) probed linear buckling and vibration problems of composite plates mixing the numerical IGA with CUF. Shishesaz et al. (2017) employed the FEM to estimate the local delamination in laminated composite plates in the post-buckling mode in the presence of a pre-central delamination defect. In addition, a multi-objective optimization study was carried out by Moradi et al. (2017) to investigate the vibrational responses of composite pipelines in the presence of magnetic field as well as considering external influences of hygro-thermal environment on the mechanical response of the system. In another numerical study, new shape functions for the Ritz-type solution were recommended by Nguyen et al. (2018) to solve both thermal buckling and vibration problems of laminated composite beams using a HSDT. Recently, the application of a control feedback system for the goal of damping the vibrational responses of laminated beams was shown by Zenkour and El-Shahrany (2019) using a HSDT to derive the motion equations of the structure.

The above discussion concerned the conventional types of composites, widely used in engineering designs. However, a novel type of composite was developed which possesses preferable properties in comparison with the well-known, previously introduced ones. In the newly developed composites, nanosize reinforcements are employed to improve the general material properties of the composite. Due to the implementation of nanosize reinforcements (i.e., small particles and/or fibers with at least one of their dimensions in the nanometers range), such composites are also known as nanocomposites. The main reason of using nanoscale elements in the nanocomposites is to gain as much as possible from the superior material properties of nanoparticles and nanofillers in comparison with their bulk specimen for the purpose of strengthening a desired matrix. The desired matrix may be metallic, polymeric, and so on. The most common type of matrix is the polymeric one, which results in generation of polymeric nanocomposites. In addition, it is noteworthy that it is possible to improve the material properties of a matrix with a wide range of nanoscale fibers or particles. One of the most famous reinforcements widely implemented for fabrication of polymeric nanocomposites is carbon nanotubes (CNTs), which can be considered as layer(s) of graphene rolled cylindrically. Both types of CNTs, namely single-walled CNTs (SWCNTs) and double-walled CNTs (DWCNTs), can be implemented to enhance the material properties of a polymeric matrix. CNTs are able to support Young's moduli of TPa order depending on their chirality. Hence, the remarkable stiffness of CNTs can improve the total stiffness of the nanocomposite properly. Due to these remarkable material properties, the mechanical analysis of *CNT-reinforced* (CNTR) nanocomposite materials and structures have attracted researchers' attention. For example, in 2000, Ajayan et al. (2000) presented a comparative study dealing with the advantages and disadvantages of SWCNTs once such nanofibers are going to be implemented as reinforcing elements in CNTR nanocomposites. Application of multi-walled CNTs (MWCNTs) as reinforcing elements in two different polymeric matrices was shown within the framework of an experimental study. It was shown that the stiffness of the enhanced nanocomposites can be dramatically increased using a low concentration of CNTs. Thostenson and Chou (2002) carried out the processing and experimental characterization of CNTR polymeric nanocomposites by emphasizing the differences between the material properties of the enriched nanocomposite whenever aligned or randomly oriented CNTs are dispersed in the initial matrix. The morphological images of the fabricated nanocomposites, viewed using a transmission electron microscope (TEM), were printed in the paper to show the manufactured nanocomposites. Gojny et al. (2004) manufactured CNTR nanocomposites reinforced with DWCNTs and measured the stiffness, strength, and fracture properties of such nanocomposite materials using a scanning electron microscope (SEM). Their comparisons reveal that the material properties obtained from the micromechanical Halpin-Tsai method can show equivalent material properties with a reliable precision. Another experimental study was undertaken by Li et al. (2004) going through mechanical characterization of SWCNTR polymeric nanocomposites. In this study, the time-dependent viscoelastic material properties of the nanocomposite (i.e., both storage and loss shear modules) were obtained by adopting a dynamic mechanical analysis (DMA) for the fabricated nanocomposite. Shi et al. (2004) dedicated their efforts to developing micromechanical methods for the purpose of presenting reliable algebraic equations to include the effects of nanotubes' waviness and agglomeration, calculating the material properties of CNTR nanocomposite materials. Goh et al. (2005) utilized the powder metallurgy technique to manufacture magnesium-based CNTR nanocomposites with maximum 0.3% w.t. CNTs and showed the enhancement in the mechanical and thermal properties of the nanocomposite. Compared with former researches, the nanocomposites made by the authors is a more efficient nanocomposite. In other words, for some CNTs, material properties of this nanocomposite are many times better than those reported in the open literature. Haque and Ramasetty (2005) developed a theoretical framework for approximation of stress state in CNTR nanocomposites with polymeric matrix. They compared their results with FE simulation to show the accuracy of their modeling. Mo et al. (2005) implemented an alumina matrix to analyze the mechanical properties of CNTR nanocomposites using SEM. It is stated in this paper that the best optimized amount for the CNTs to be dispersed in the matrix is 1.5% w.t. and after this amount the fracture toughness will be decreased. Furthermore, Rajoria and Jalili (2005) surveyed the damped vibrational characteristics of CNTR nanocomposites using both SWCNTs and MWCNTs to strengthen the epoxy matrix. They realized that the damping properties of the nanocomposite are much better than those of the pure matrix material. In another paper, the viscoelastic analysis of nanocomposites reinforced with CNTs was performed by Suhr et al. (2005). Results of this article reveal an improvement in the damping behaviors of nanocomposites with a finite amount of CNTs scattered in the polymeric matrix. Afterward, Kim et al. (2006) fabricated CNTR Cu-based nanocomposites to measure the stress and strength characteristics of such materials. The nanocomposite was processed using spark plasma sintering (SPS) followed by cold rolling. Using SEM, the presented graphs reveal the optical micrographs of CNTR Cu-based nanocomposites with different amounts of CNTs percentage. Seidel and Lagoudas (2006) presented a group of various micromechanical frameworks to approximate the material properties of CNTR nanocomposites. Investigation of the large strain constitutive behaviors of CNTR nanocomposite materials was carried out by Tan et al. (2007), based upon the nonlinear cohesive law regarding for van der Waals (vdW)-based interface. The reverse destroying impact of adding CNTs to the initial matrix in high strains is covered in this study. Experimental research authored by Xiao et al. (2007) considers both mechanical and rheological characteristics of CNTR polymeric nanocomposites. In this paper, low density polyethylene (LDPE) nanocomposites are utilized to procure the dynamic tests at various test frequencies. It can be seen that the viscoelastic responses of the material can be changed by tuning the test frequency depending on the amount of CNT employed to reinforce the polymeric matrix. Zhu et al. (2007) performed a molecular dynamics (MD) simulation dealing with the constitutive behaviors of CNTR nanocomposites manufactured utilizing either short or long SWCNTs. They compared the results with those enriched from the rule of the mixture and it was shown that up to a critical strain, the simulations are in complete agreement with those obtained from the rule of the mixture. Tserpes et al. (2008) presented FE-based multi-scale modeling to estimate the tensile properties of CNTR nanocomposites using a representative volume element (RVE). In another experimental study, both electrical and mechanical properties of copper-based CNTR nanocomposites were examined (Daoush et al., 2009). Kim et al. (2009) carried out an experimental study dealing with manufacturing, characterization, and modeling of a new type of CNTR nanocomposites which consist of carbon fibers as well as nanosize CNTs. It was reported that the mechanical performance of such three-phase nanocomposites are better than other types. The influences of nanofibers' debonding and wavy nature were included in the framework of a micromechanical investigation proposed by Shao et al. (2009). Formica et al. (2010) produced an FE-based investigation dealing with the dynamic responses of CNTR nanocomposite plates. The equivalent material properties were enriched using the well-known Eshelby-Mori-Tanaka method. On the other hand, first-order shear deformation theory (FSDT) of beam-type elements was employed by Ke et al. (2010) to probe the natural frequency characteristics of functionally graded CNTR (FG-CNTR) nanocomposite beams considering the effects of nonlinear strain-displacement relationships on the frequency of the nanocomposite structure. Shen and Zhang (2010) considered the impacts of geometrical imperfection of the plates while analyzing thermal postbuckling behaviors of FG-CNTR nanocomposite structures. The influences of various approximated effective terms in determining the material properties of CNTR nanocomposites were studied by Bakshi and Agarwal (2011), dealing with the aluminum-based nanocomposites. It was shown that micromechanical schemes such as rule of the mixture and Halpin-Tsai models are valid for limited ranges of CNTs' volume fraction. A plasticity-based framework was developed by Barai and Weng (2011) for the goal of reaching the stress-strain curves of elastoplastic CNTR nanocomposite. The thermo-mechanical stability analysis of FG-CNTR nanocomposite shells was produced by Shen (2012). The application of the Eshelby-Mori-Tanaka homogenization method for the vibration problem of FG-CNTR nanocomposite panels was shown by Sobhani Aragh et al. (2012).

They solved the final motion equations in the framework of the numerical generalized differential quadrature method (GDQM). Also, the Timoshenko beam hypothesis was implemented by Yas and Samadi (2012) for the purpose of reaching both stability and frequency responses of FG-CNTR nanocomposite beams rested on elastic substrate. Similarly, Zhu et al. (2012) devoted their article to probe bending, buckling, and vibration problems of FG-CNTR nanocomposite plates on the basis of the FSDT of plates combined with the FEM. Alibeigloo and Liew (2013) presented a three-dimensional (3-D) elasticity solution for thermal bending responses of CNTR nanocomposite plates. In another study, Ke et al. (2013) probed the dynamic buckling analysis of FG-CNTR nanocomposite beams using the well-known Timoshenko beam model. The investigation of buckling and thermal vibration problems of FG-CNTR nanocomposite plates was performed by researchers on the basis of element-free kernel particle Ritz (kp-Ritz) FE-based discretization (Lei et al., 2013a; Lei et al., 2013b). Large amplitude free vibration and thermally influenced buckling responses of smart piezoelectric CNTR nanocomposite beams were explored by Rafiee et al. (2013a) and Rafiee et al. (2013b) respectively, according to the nonlinear von Karman expansion of the strain-displacement relations of Euler-Bernoulli beams. According to an experimental framework, Tehrani et al. (2013) added MWCNTs to FRCs to monitor the probable changes in the mechanical behaviors of such new nanocomposites. A group of material properties like ultimate strength, tensile characteristics, fracture strain, storage modulus, and loss factors of this type of CNTR nanocomposites were surveyed. This paper reports an increase in the fracture strain of CNTR-FRCs in comparison with simple FRCs without CNTs, whereas the ultimate limit and elastic moduli of these two types are not away from each other. A HSDT was employed by Wattanasakulpong and Ungbhakorn (2013) for the goal of enriching deflection, buckling load, and natural frequency of nanocomposite beams reinforced with CNTs. The well-known GDQM was implemented by Ansari et al. (2014) to extract the natural frequency responses of FG-CNTR nanocomposite beams. They derived the beam's motion equations based upon the Timoshenko beam hypothesis. Arash et al. (2014) implemented MD simulations to examine the fracture behaviors of CNTR nanocomposite materials. They also extended micromechanical relations for deriving the moduli of these nanocomposites and compared their simulation with the proposed theoretical procedure. It was shown that a reliable agreement can be observed between these two approaches. Vibrational characteristics of FG-CNTR nanocomposite conical shells were probed by Heydarpour et al. (2014) according to the expansion of the linear displacement field of FSDT of shells for rotating conical shells. The achieved governing equations of motion were solved on the basis of the DQM. The well-known Eshelby-Mori-Tanaka micromechanical scheme was employed by Lei et al. (2014) to derive the material properties of FG-CNTR nanocomposites to analyze the dynamic stability behaviors of panels fabricated from this material. They utilized an element-free kp-Ritz discretization method to derive the final answer of the problem. Similarly, another FE-based solution method was implemented by Liew et al. (2014) to find the postbuckling responses of FG-CNTR nanocomposite panels whenever the structure is subjected to an axial compression. Both first- and third-order kinematic beam hypotheses were used by Lin and Xiang (2014) to reach the natural frequency of FG-CNTR nanocomposite beams. Malekzadeh and Zarei (2014) surveyed dynamic characteristics of FG-CNTR nanocomposite quadrilateral plates. They employed mapping technique to solve the final derived partial differential equations (PDEs) of the problem. Both stiffness and fracture characteristics of CNTR nanocomposites were investigated by Wernik and Meguid (2014) using experimental tests. Static and dynamic characteristics of FG-CNTR nanocomposite panels were investigated on the basis of the Eshelby-Mori-Tanaka micromechanical procedure in association with the FSDT of panels (Zhang et al., 2014a; Zhang et al., 2014b). The study of forced vibration problem of FG-CNTR nanocomposite plates was performed by Ansari et al. (2015) using the efficient GDQ numerical method. In this analysis, impacts of shear deformation are included employing FSDT. Another study dealing with mechanical stability of FG-CNTR nanocomposite conical shells was carried out by Jam and Kiani (2015a) regarding the external pressure applied on the nanocomposite structure. In addition to the previous paper, the same authors produced a low-velocity impact analysis about FG-CNTR nanocomposite beams once the continuous system is subjected to thermal loading (Jam and Kiani, 2015b). An IGA was conducted by Phung-Van et al. (2015) to probe both deflection and natural frequency responses of FG-CNTR nanocomposite plates using HSDT of plates. They verified the validity of their extracted answers, comparing them with those achieved from FEM. Zhang et al. (2015a) developed an IMLS-Ritz method for solving the vibration problem of FG-CNTR nanocomposite triangular plates within the framework of the FSDT incorporated with the well-known rule of the mixture. They could reach the vibrational responses of the plate based upon an element-free process. The same steps were followed by Zhang et al. (2015b) for the purpose of reaching natural frequency of FG-CNTR nanocomposite skew plates. In another endeavor, the von Karman strain-displacement relations of the Kirchhoff-Love plate model were considered by Ansari et al. (2016) to derive the nonlinear postbuckling answers of FG-CNTR nanocomposite plates on the basis of an analytical solution. Stability and frequency behaviors of FG-CNTR nanocomposite conical shells were covered in an analysis performed by Ansari and Torabi (2016) regarding for axial loading applied on the nanocomposite structure. Also, Mirzaei and Kiani (2016) implemented Chebyshev polynomial approximation functions to solve the free vibration problem of FG-CNTR nanocomposite cylindrical panels according to the well-known FE-based Ritz method. They derived the Euler-Lagrange equations of the structure based on the FSDT. The influences of geometrical imperfection of the beam were covered by Wu et al. (2016) investigating the nonlinear frequency behaviors of FG-CNTR nanocomposite beams. The natural frequency responses were derived using a new representation of the Ritz-type FEM. Another continuum mechanics-based study was procured by Fantuzzi et al. (2017) to derive the natural frequencies of FG-CNTR nanocomposite plates with arbitrary shapes. They founded their research on the concept of IGA to support various shapes for the structure.

   On the other hand, it must be noted that CNTs are not the only nanosize reinforcing elements used to strengthen initial matrices. In fact, other carbonic nanoparticles and nanofibers can be employed in the fabrication of the nanocomposite materials. For example, graphene, graphene platelet (GPL), and graphene oxide (GO) are other common types of carbon-based nanomaterials that can reinforce an initial matrix. For the sake of simplicity, nanocomposites made by the means of these reinforcements are called *graphene-reinforced* (GR) nanocomposites, *GPL-reinforced* (GPLR) nanocomposites, and *GO-reinforced* (GOR) nanocomposites, respectively. Several research projects can be observed in the scientific society dealing with the performance of such nanocomposites in different working conditions which resulted in publication of a large number of papers in valuable journals. The impacts of the amount of GPLs used in the composition of hybrid CNT/GPL/polymer nanocomposites on the fracture toughness, stiffness, and thermal conductivity of such nanocomposites were investigated by Chatterjee et al. (2012a) regarding the effects of GPLs' dimensions on the mechanical performance of the polymeric nanocomposite. In another paper, the influence of using amine functionalized GPLs in manufacturing GPLR polymeric nanocomposites to improve mechanical and thermal properties of this type of nanocomposites was investigated based on an experimental study conducted by Chatterjee et al. (2012b) using Raman spectroscopy for the purpose of characterizing the fabricated nanocomposite. Additional experimental research was performed by Li et al. (2013), exploring the influences of simultaneously employing CNTs and GPLs to fabricate hybrid nanocomposite materials on the mechanical behaviors of such nanocomposites. It was shown that such nanocomposites can provide improved mechanical properties in comparison with neat epoxy or nanocomposites reinforced with either CNT or GPL. Nieto et al. (2013) manufactured GPLR tantalum carbide (TaC)-based nanocomposites and studied the effects of adding GPLs on the fracture toughness of the fabricated nanocomposite. Yadav and Cho (2013) carried out a comparative experimental investigation dealing with the effect of using functionalized GPLs instead of non-functionalized ones to determine the material properties of GPLR nanocomposite materials. In another experimental study, it was determined that hybrid GPLR nanocomposites that possess carbon fibers can provide the worst working conditions because of the enhancement of the mechanical and thermal properties of such nanocomposites compared with neat polymer, composites reinforced with carbon fibers, and nanocomposites reinforced with GPLs (Yang et al., 2013). Chandrasekaran et al. (2014) examined the impact of using various carbonic nanofillers on the fracture toughness of nanocomposites. They employed CNTs, GPLs, and reduced GOs as reinforcing elements and observed that the best choice for the purpose of increasing the fracture toughness is GO. Furthermore, Naebe et al. (2014) showed the influence of implementing functionalized GPLs to increase the viscoelastic mechanical properties of GPLR nanocomposites. Pérez-Bustamante et al. (2014) showed the effect of adding GPLs to aluminum matrix on the hardness characteristics of GPLR nanocomposites. It was shown that the longer is the milling time, the harder will be the achieved nanocomposite. The SPS method was employed by Shin and Hong (2014) to make GOR nanocomposites using a yttria-stabilized zirconia matrix. Thereafter, the electrical, thermal, and mechanical properties of the manufactured nanocomposite were derived according to an experimental characterization. Stiffness and fatigue behaviors of hybrid nanocomposites reinforced with combination of GPLs and nanoscale carbon fibers were investigated in an experimental study carried out by Shokrieh et al. (2014a). In addition, Shokrieh et al. (2014b) studied the impacts of using GPLs and graphene nanosheets (GNSs) on the fracture toughness and tensile strength of hybrid nanocomposites. Wu and Drzal (2014) tried to extract the coefficient of thermal expansion of GPLR nanocomposites experimentally. They captured the effects of the annealing procedure on the variation of the CTE of the nanocomposite. The influence of simultaneous implementation of CNTs and GPLs for the purpose of reinforcing epoxy matrix on the electro-mechanical properties of polymeric nanocomposites was explored by Yue et al. (2014). Another experimental study was undertaken by Ahmadi-Moghadam et al. (2015) concerning with the effect of functionalization of GPL on the fracture toughness of GPLR nanocomposite materials. Both functionalized and non-functionalized GPLs were utilized by Ahmadi-Moghadam and Taheri (2015) to analyze the effects of nanofillers' functionalization on the fracture behaviors of such nanocomposites. All three modes of fracture were covered within this experimental investigation for GPLR nanocomposite materials. The mechanical characterization of FRC-based GPLR nanocomposites was carried out by Hadden et al. (2015) based on an MD simulation validated by experimental tests. Kamar et al. (2015) analyzed the consequences of adding GPLs to epoxy composites in order to monitor the probable changes in the interlaminar behaviors of such GPLR nanocomposites. They observed that the best choice is to add 0.25 w.t. GPLs to the matrix to possess greatest flexural moduli and mode-I fracture toughness, simultaneously. Rashad et al. (2015) implemented GPLs to fabricate metallic nanocomposites based on a magnesium matrix to develop better material properties at the room temperature. The fabrication procedure was completed using the powder metallurgy technique and the characterization was fulfilled based on the microscopic observations in the nanoscale. Mechanical, thermal, and viscoelastic properties of GPLR nanocomposites with 3%–5% volume fractions were reported in an article proposed by Wang et al. (2015) in an experimental manner. The issue of probing tensile properties of GPLR nanocomposites was the main objective of an experimental study conducted by Liang et al. (2016). Rashad et al. (2016) reported an increase in the hardness and stiffness of Mg-6Zn alloys by adding a limited amount of nanosize GPLs to the initial material. They accomplished the tests once they fabricated the specimen via disintegrated melt deposition method. The yield strength, ductility, and hardness of aluminum-based GPLR nanocomposites were explored in an investigation conducted by Bisht et al. (2017), who manufactured the specimens by SPS method. They showed that the best improvement can be achieved using 1 w.t. for the GPLs in the composition of the nanocomposite.

As stated in the above-studied literature, the mechanical properties of polymeric and/or metallic matrices can be enhanced when a proper amount of nanosize reinforcements are added to them. Investigation of the reaction of structures fabricated from such advanced materials once they are subjected to either static or dynamic excitations is one of the hot topics in engineering. Therefore, some of the researchers tried to provide a better understanding of the mechanical behaviors of nanocomposite structures reinforced with carbonic nanomaterials excluding CNTs. One of the primary attempts in this area includes the nonlinearity effects when studying bending and free vibration problems of polymeric nanocomposite beams reinforced with GPLs (Feng et al., 2017a; Feng et al., 2017b). Kitipornchai et al. (2017) used FSDT to analyze the natural frequency and buckling load characteristics of GPLR nanocomposite beams once there exists porosities in the nanocomposite material.

The nonlinear large deflection characteristics of GPLR nanocomposite plates were surveyed by Gholami and Ansari (2017) in the framework of the sinusoidal shear deformation plate hypothesis using the well-known numerical GDQM. Shen et al. (2017a) conducted a vibration study dealing with the nonlinear behaviors of GR nanocomposite beams once the nanocomposite is embedded on an elastic foundation and it is placed in a thermally influenced environment. In another paper, the issue of thermo-mechanical stability behaviors of GR nanocomposite plates was investigated by Shen et al. (2017b) based upon the third-order plate theory of Reddy. In another study, the Mindlin-Reissner kinematic plate theory was implemented by Song et al. (2017a) to probe both free and forced frequency responses of GPLR nanocomposite structures. Furthermore, the nonlinear buckling responses of GPLR nanocomposite multi-layered plates were extracted on the basis of the FSDT by Song et al. (2017b) using a two-step perturbation technique. The DQM was utilized by Wu et al. (2017b) in association with the strain-displacement relations of the Timoshenko beam hypothesis to solve the dynamic buckling problem of GPLR nanocomposite beam-type structures. A refined shear deformable beam hypothesis was employed by Barati and Zenkour (2017) to solve the nonlinear postbuckling problem of porous GPLR nanocomposite beams while the impacts of initial geometrical imperfection are included. Some of the researchers presented a 3-D elasticity solution for the thermally affected bending responses of GPLR nanocomposite plates, either rectangular or circular/annular (Yang et al., 2017a; Yang et al., 2017b). The nonlinear stability characteristics of GPLR nanocomposite beams were studied by Yang et al. (2017c) on the basis of the Timoshenko beam model. The reaction of trapezoidal GPLR nanocomposite structures to bending and vibrating excitations was reported by Zhao et al. (2017) with the aid of an FE-based commercial software. In hybrid, novel research, the effect of adding GPLs to CNTR nanocomposites on the bending and vibration behaviors of nanocomposite plates was investigated by García-Macías et al. (2018). The dynamic reaction of GPLR nanocomposite plates to the harmonic external excitation was monitored in an investigation completed by Gholami and Ansari (2018) according to the Reddy's plate model incorporated with the variational DQM (VDQM). The vibrational behaviors of GPLR nanocomposite quadrilateral plates were studied by Guo et al. (2018) using an FE-based meshfree method, called IMLS-Ritz method. In addition, a 3D elasticity solution was developed by Liu et al. (2018) for solving buckling and vibration problems of GPLR nanocomposite cylinders while an initial pre-stress exists in the structure. A general thermo-mechanical analysis was accomplished by Rafiee et al. (2018) dealing with bending, buckling, and vibration problems of hybrid FRC-based GPLR nanocomposites. In another effort, the axially compressed thermal stability problem of GR nanocomposite panels was solved by Shen et al. (2018a) using the third-order plate model proposed by Reddy. Continuing with the analysis of the mechanical responses of GR nanocomposite panels, the thermo-elastic nonlinear frequency responses of such structures resting on an elastic medium were derived by Shen et al. (2018b), too. The deflection and stability analyses of GPLR nanocomposite plates were performed by Song et al. (2018) according to the FSDT. Wang et al. (2018a) analyzed bending and vibrating responses of GPLR nanocomposite doubly-curved shells using a HSDT. They showed the accuracy of their modeling by comparing their results with those reported in the open literature and achieved employing the FE-based ANSYS commercial software. A group of researchers used FEM to derive the buckling load behaviors of GPLR nanocomposite cylinders with cutout (Wang et al., 2018b; Wang et al., 2018c). The dynamic stability responses of GPLR nanocomposite plates were reported in an article conducted by Wu et al. (2018). A 3D elasticity framework was implemented by Yang et al. (2018) to derive the stress and deflection responses of GPLR nanocomposite elliptical plates whenever the constitutive equations of the nanocomposite treat as the same those of transversely isotropic materials. The bending deflection, buckling load, and natural frequency responses of GOR nanocomposite beams were extracted by Zhang et al. (2018) using a variational-based energy method incorporated with the Timoshenko beam model. Later, the HSDT of beams was used by Ebrahimi et al. (2019a) to consider the distribution of shear stress and strain while probing the vibrational responses of GOR nanocomposite beam-type structures in the presence of an external magnetic field. Moreover, the thermo-elastic frequency analysis of GOR nanocomposite plates was produced by Ebrahimi et al. (2019c) once the structure is rested on a two-parameter elastic medium. The issue of analyzing thermal buckling responses of GPLR nanocomposite plates was surveyed by Ebrahimi and Qaderi (2019) using the Reddy's plate hypothesis. Moreover, Qaderi et al. (2019) explored the damped vibrational behaviors of GPLR nanocomposite beams utilizing a HSDT. In other research, a comparative study was accomplished by Rafiee et al. (2019) to investigate the best choice for the purpose of improving the thermal conductivity of nanocomposite materials. Lately, both static buckling and natural frequency analyses of GOR nanocomposite plates were accomplished considering the effects of various terms, including resting media and thermal loading, on the mechanical response of the nanocomposite structure (Ebrahimi et al., 2019b; Ebrahimi et al., 2019d).

1.2 APPLICATIONS OF NANOCOMPOSITES IN INDUSTRY

Here we will pay attention to the most significant applications of nanocomposites in modern engineering designs. As stated in the previous paragraphs, nanomaterials enriched by dispersing nanosize particles and fillers can provide better mechanical, thermal, and electrical properties compared to the homogeneous matrix itself. Obviously, the cost of an employed material is one of the determining parameters in its selection to be used in the industrial applications as well as its performance. Some of the nanocomposite materials exhibit a very high cost-to-performance ratio which satisfies one of the most crucial requirements of the design.

Nobody doubts that one of the most important industries in the world is the automotive industry. This industry is in direct contact with the people's lives as well as with the environment. All of the efforts made in this area aim to satisfy four important criteria: first, to provide a more comfortable atmosphere for the customers to increase the sense of pleasure which will induced them to buy the automobiles; second, to lessen extra costs for the costumers by reducing the fuel costs, third, to help save environmental resources by lessening the amount of the fuel which must be used by the automobile, and fourth, to produce high-technology automobiles that are safe for the customer. Due to these criteria, one of the industries that is moving

toward implementation of nanocomposite materials is surely the automotive industry. The reason is clear: nanocomposite materials exhibit at least similar behavior
to that of metallic materials. In automotive applications, polymer-clay nanocomposites are widely used as the constituent material to manufacture various metallic
parts of an automobile. Indeed, this helps to reach better mechanical performance
in the presence of cost saving as well as weight saving, which is another important
parameters in any engineering design. Some concepts to reduce the weight of the
automotive devices neglect the superior material properties of the nanocomposites and attempt to implement plastics, glass fiber-reinforced composites, and light-
weight alloys of metals; however, each of these alternatives results in increased
fabrication costs. However, adding a limited amount of the nanosize reinforcements
in a host material can be resulted in the improvement of the stiffness, toughness,
thermal stability, fire retardancy, and surface hardness via a low amount of cost.
One of the best choices in this area is to utilize nanoclay for the purpose of reinforcing the polymeric matrices to be used in the automotive devices. The advantage
of this type of the polymeric nanocomposite is the simple fabrication which can be
completed by melt compounding or the conventional solution method. Following
this concept, the Toyota Motors Corporation used nanoclay polymeric nanocomposites to produce the timing-belt of some of its cars (Garcés et al., 2000), resulting
in a dual cost saving. In other words, the cost saved by avoiding implementation
of metallic materials is the first saving followed by the lower fuel spent due to the
reduction in the weight of the car.

On the other hand, it must be pointed out that producing nanocomposites that
are efficient enough to be used in the automotive industry requires a critical con-
sideration of a large number of requirements. For instance, the superior mechanical, thermal, and electrical properties of the polymeric nanocomposites can only
be achieved by employing tiny, nanosize reinforcements in the host matrix. In fact,
the dimensions of the reinforcement gadget is one of the key factors in determining
the final stiffness that can be achieved by fabricating such a nanomaterial. In other
words, high volume-to-surface ratio of the nanoparticles and/or nanofillers is one
of the most crucial terms that can affect the final equivalent stiffness. Hence, only
nanocomposites fabricated from nanosize reinforcements with high volume-to-surface
ratio can meet the desired design purposes. It might be interesting to know that the
polymeric nanocomposite that was manufactured by Toyota Motors Corporation in
collaboration with Ube Industries was able to provide stiffness of nearly twice that
of the initial matrix itself (Garcés et al., 2000). The strengthening was achieved by
using nanoclay to the extent of only 1.6 percent of the material's entire volume. This
is the result of the high volume-to-surface ratio of the nanoclay utilized in the fabrication of the nanocomposite material.

Because of the simple fabrication of the polyolefin-based nanocomposite materials, the act of manufacturing and probing the performance of such materials is of high
significance to both industrial and academic communities. It is reported that such
nanocomposites can be manufactured by manipulating the catalysts. Furthermore,
polypropylene (PP)-based nanocomposites are excellent alternatives due to the simple fabrication of the PP via well-known fabrication procedures containing extru-
sion. At the Toyota Motor Corporation R&D center, nanoclay nanocomposites were

manufactured using the modified PP as the host material. They resulted in much stiffer materials compared to the PP itself (Garcés et al., 2000).

On the other hand, the automotive industry is not the only one concerned with implementing nanocomposite materials. There are many other industries that are involved with such implementation; however, all of them will not fall into the scope of mechanical engineering and related sciences. As one of the most relevant applications of nanocomposites in the field of mechanics, the aerospace applications can be mentioned. In this strategic industry, polymeric nanocomposites are of great importance due to their particular material properties. These nanocomposites can show improved stiffness, electrical and thermal conductivity, fire retardancy, and so on. For example, clay nanocomposites can reveal preferable corrosion resistance, fire retardancy, thermal stability, and impact resistance which all are of great importance in the aerospace industry. Also, CNTR polymeric nanocomposites can show remarkable thermal stability and impact resistance as well as great stiffness. It is noteworthy that in the polymeric nanocomposites, the interphase which will appear due to the remarkable difference in the material properties of the nanofiller and host matrix is one of the key characters that causes the widespread application of such nanomaterials in the aerospace industry. A very obvious desirable design criterion is the lower weight of the designed device. This design criterion is more important in the aerospace designs concerned with designing aircrafts and spacecraft. Due to this, the Boeing Company increased the proportion of the composites and nanocomposites in its newer designs for its products. The discussion in this paragraph was intended to give a general insight to the readers about the crucial applications of the nanocomposites in this industrial area. However, more detailed data can be found by referring to the complementary references (Joshi and Chatterjee, 2016; Kausar et al., 2017).

2 Mathematical Tools

2.1 AN INTRODUCTION TO ELASTICITY

This section will go through the derivation of the strain tensor of the continua in the framework of the elasticity preliminary relations. The final relation will enable the reader to obtain the components of the strain tensor for any desired continuous system. It is presumed that readers are skilled enough to manipulate the algebraic operations between various tensors and matrices. Therefore, no preliminary algebraic review will be presented and those who are not familiar with these concepts are asked to strengthen their mathematical understanding before reading what follows.

Assume two points P and Q placed in a continuous system with a desired distance $d\mathbf{X}$. It is assumed that the secondary state of such points under actuation of the deformation gradient tensor $\mathbf{F}$ will be P' and Q' with algebraic distance $d\mathbf{x}$. In fact, vectors $\mathbf{X}$ and $\mathbf{x}$ denote the displacement vectors in the undeformed and deformed coordinates respectively under action of the deformation gradient tensor $\mathbf{F}$. Hence, the displacement between points P and Q can be defined in the following form:

$$(dS)^2 = d\mathbf{X}.d\mathbf{X} \tag{2.1}$$

Similarly, the displacement between points P' and Q' in the deformed situation can be expressed as follows:

$$(ds)^2 = d\mathbf{x}.d\mathbf{x} \tag{2.2}$$

Using the well-known relationship between deformed and undeformed displacement vectors in the continuum mechanics ($\mathbf{x} = \mathbf{F}.\mathbf{X}$), Eq. (2.2) can be rewritten as follows:

$$(ds)^2 = \left[d\mathbf{X}.\mathbf{F}^{\mathrm{T}}\right].\left[\mathbf{F}.d\mathbf{X}\right] = d\mathbf{X}.\left[\mathbf{F}^{\mathrm{T}}.\mathbf{F}\right].d\mathbf{X} \tag{2.3}$$

where $\mathbf{C} = \mathbf{F}^{\mathrm{T}}.\mathbf{F}$ is the *right Cauchy-Green deformation* tensor. Therefore, the displacement in the deformed domain can be derived using the following relation:

$$(ds)^2 = d\mathbf{X}.\mathbf{C}.d\mathbf{X} \tag{2.4}$$

On the other hand, the *left Cauchy-Green deformation* tensor can be defined as follows:

$$\mathbf{B} = \mathbf{F}.\mathbf{F}^{\mathrm{T}} \tag{2.5}$$

It is obvious that due to the symmetrical nature of the deformation gradient tensor, $\mathbf{F}$, both right and left Cauchy-Green deformation tensors are symmetric tensors.

The deformation of a continuous system, produced by the deformation gradient tensor, results in a difference between the squares of displacements in the reference and current coordinates. Subtracting these two squares from each other results in reaching the below relation:

$$(ds)^2 - (dS)^2 = 2d\mathbf{X}.\mathbf{E}.d\mathbf{X} \tag{2.6}$$

in which $\mathbf{E}$ is the *Lagrangian* or *Green-Lagrange* strain tensor and can be defined in the following form:

$$\mathbf{E} = \frac{1}{2}(\mathbf{C} - \mathbf{I}) = \frac{1}{2}(\mathbf{F}^T.\mathbf{F} - \mathbf{I}) \tag{2.7}$$

Regarding the relationship between the deformation gradient tensor and gradient operation with respect to the reference coordination, the aforementioned relation can be modified as follows:

$$\begin{aligned}\mathbf{E} &= \frac{1}{2}\left([\mathbf{I} + \nabla_0\mathbf{u}].[\mathbf{I} + \nabla_0\mathbf{u}]^T - \mathbf{I}\right) \\ &= \frac{1}{2}\left(\nabla_0\mathbf{u} + [\nabla_0\mathbf{u}]^T + \nabla_0\mathbf{u}.[\nabla_0\mathbf{u}]^T\right)\end{aligned} \tag{2.8}$$

It is worth mentioning that including the terms generated by $\nabla_0\mathbf{u}.[\nabla_0\mathbf{u}]^T$ results in reaching the nonzero strain-displacement relations of the continua. Now, to give insight about the calculations required to reach the components of the Lagrangian strain tensor, the index notation will be utilized. In other words, the *IJ* component of the Lagrangian strain tensor can be achieved using the following relationship:

$$E_{IJ} = \frac{1}{2}\left(\frac{\partial u_I}{\partial X_J} + \frac{\partial u_J}{\partial X_I} + \frac{\partial u_K}{\partial X_I}\frac{\partial u_K}{\partial X_J}\right) \tag{2.9}$$

In above equation, all of the differentiations are computed with respect to reference configuration because of the fact that the gradient operator was employed based upon the reference configuration. Dismissing the nonlinear terms in above relation, the components of the linearized strain tensor can be achieved using the following equation:

$$E_{IJ} = \frac{1}{2}\left(\frac{\partial u_I}{\partial X_J} + \frac{\partial u_J}{\partial X_I}\right) \tag{2.10}$$

The above discussion was concerned with derivation of the strain tensor in the framework of Lagrangian description. Now, the squares of displacements in the deformed and undeformed conditions will be again subtracted to derive the strain tensor in the *Eulerian* description. In other words, we will try to present a mathematical equivalent expression for dS in terms of ds instead of that expressed in the former manipulation. Indeed, the displacement between P and Q points can be expressed by:

$$(dS)^2 = \underbrace{d\mathbf{x}.\mathbf{F}^{-T}}_{d\mathbf{X}}.\underbrace{\mathbf{F}^{-1}.d\mathbf{x}}_{d\mathbf{X}} = d\mathbf{x}.\left[\mathbf{F}^{-T}.\mathbf{F}^{-1}\right].d\mathbf{x} \qquad (2.11)$$

where

$$\tilde{\mathbf{B}} = \mathbf{F}^{-T}.\mathbf{F}^{-1} \qquad (2.12)$$

is the *Cauchy* strain tensor; now, using the definition of the Cauchy strain tensor incorporated with Eq. (2.11), the displacement in the reference configuration can be extracted using the following relation:

$$(dS)^2 = d\mathbf{x}.\tilde{\mathbf{B}}.d\mathbf{x} \qquad (2.13)$$

Now, the subtract of squares of displacements in the reference and current configurations can be rewritten as follows:

$$(ds)^2 - (dS)^2 = 2d\mathbf{x}.\mathbf{e}.d\mathbf{x} \qquad (2.14)$$

where $\mathbf{e}$ is the *Eulerian* or *Almansi* strain tensor. This tensor can be expressed by the following equation:

$$\mathbf{e} = \frac{1}{2}\left(\mathbf{I} - \tilde{\mathbf{B}}\right) = \frac{1}{2}\left(\mathbf{I} - \mathbf{F}^{-T}.\mathbf{F}^{-1}\right) \qquad (2.15)$$

Now, the deformation gradient tensor will be considered to be a function of identity tensor and gradient operator in the current configuration. So, the Eulerian strain tensor can be defined as:

$$\begin{aligned}\mathbf{e} &= \frac{1}{2}\left(\mathbf{I} - [\mathbf{I} - \nabla\mathbf{u}].[\mathbf{I} - \nabla\mathbf{u}]^T\right) \\ &= \frac{1}{2}\left(\nabla\mathbf{u} + [\nabla\mathbf{u}]^T - \nabla\mathbf{u}.[\nabla\mathbf{u}]^T\right)\end{aligned} \qquad (2.16)$$

Like Eq. (2.8), Eq. (2.16) can be presented using index notation to reach the components of the Eulerian strain tensor. Doing the mathematical manipulations, the arrays of Eulerian strain tensor can be computed in the following form:

$$e_{ij} = \frac{1}{2}\left(\frac{\partial u_i}{\partial x_j} + \frac{\partial u_j}{\partial x_i} - \frac{\partial u_k}{\partial x_i}\frac{\partial u_k}{\partial x_j}\right) \qquad (2.17)$$

Ignoring the nonlinear terms generated in the Eq. (2.17), the linearized Eulerian strain tensor can be defined in the following form:

$$e_{ij} = \frac{1}{2}\left(\frac{\partial u_i}{\partial x_j} + \frac{\partial u_j}{\partial x_i}\right) \qquad (2.18)$$

In what follows, Eq. (2.18) will be implemented to derive the strain-displacement relations of any desired continuous system which is assumed to be analyzed in the linear domain.

2.2 KINEMATIC RELATIONS OF CONTINUOUS SYSTEMS

In this section we investigate various kinematic hypotheses related to derive the motion equations of continuous systems. In this book, shells, beams, and plates will be analyzed once they are subjected to either static or dynamic excitations. Therefore, it is necessary to discuss the motion equations of such elements while they are subjected to various types of external loadings. In what follows, first-order cylindrical shell theory, first-order beam theory, third-order beam theory, refined higher-order beam theory, third-order plate theory, and refined higher-order plate theory will be utilized. So, the derivation of the motion equations of continuous systems simulated via each of these theories will be expressed in what follows. All of the motion equations will be derived for the dynamic form first. In other words, the influence of time will be included in the derivation of the motion equations to achieve a general set of governing equations for the dynamic problems. When the static stability analysis is going to be carried out, the terms that possess differentiation with respect to the time will be dismissed to enrich the motion equations in the static state. Moreover, because the equations of motion are going to be derived for the dynamic form, the displacement fields of the continuous systems will be assumed to be time-dependent fields; however, this time-dependency will disappear in the case of surveying static buckling problems.

2.2.1 CLASSICAL SHELL THEORY

In this section the mathematical representation of classical shell theory, also known as Donnell shell theory, will be presented. According to this theorem, the influences of the shear stress and strain will be dismissed due to their small effects in comparison with those related to the bending mode. Even though this theory cannot cover shear deformations, it can provide an accurate response with an acceptable precision for thin-walled shells with great radius-to-thickness ratios. Hence, we are about to present a mathematical framework for thin cylindrical shells within the framework of this theory. Based on this model, the displacement field of the shell can be presented in the following form:

$$u_x(x,\varphi,z,t) = u(x,\varphi,t) - z\frac{\partial w(x,\varphi,t)}{\partial x},$$

$$u_\varphi(x,\varphi,z,t) = v(x,\varphi,t) - z\frac{\partial w(x,\varphi,t)}{\partial \varphi}, \tag{2.19}$$

$$u_z(x,\varphi,z,t) = w(x,\varphi,t)$$

in which u, v, and w are the axial displacement, circumferential displacement, and bending deflection, respectively. Now, the strain-displacement relations of this type of thin shell can be derived on the basis of the former relations derived for the strain tensor of a continuum. Considering the nonlinear strains of a continuous system (see Eq. (2.17)), the components of the strain tensor can be derived as follows:

$$\varepsilon_{xx} = \frac{1}{2}\left(\frac{\partial u_x}{\partial x} + \frac{\partial u_x}{\partial x} - \frac{\partial u_z}{\partial x}\frac{\partial u_z}{\partial x}\right) = \frac{\partial u}{\partial x} + \frac{1}{2}\left(\frac{\partial w}{\partial x}\right)^2 - z\frac{\partial^2 w}{\partial x^2},$$

$$\varepsilon_{\varphi\varphi} = \frac{1}{2}\left(\frac{\partial u_\varphi}{\partial \varphi} + \frac{\partial u_\varphi}{\partial \varphi} - \frac{\partial u_z}{\partial \varphi}\frac{\partial u_z}{\partial \varphi}\right) = \frac{1}{R}\left(\frac{\partial v}{\partial \varphi} + w + \frac{1}{2}\left(\frac{\partial w}{\partial \varphi}\right)^2\right) - \frac{z}{R^2}\left(\frac{\partial^2 w}{\partial \varphi^2} - \frac{\partial v}{\partial \varphi}\right), \quad (2.20)$$

$$\varepsilon_{x\varphi} = \frac{1}{2}\left(\frac{\partial u_x}{\partial \varphi} + \frac{\partial u_\varphi}{\partial x} - \frac{\partial u_z}{\partial x}\frac{\partial u_z}{\partial \varphi}\right) = \frac{1}{R}\frac{\partial u}{\partial \varphi} + \frac{\partial v}{\partial x} + \frac{1}{R}\frac{\partial w}{\partial x}\frac{\partial w}{\partial \varphi} - \frac{z}{R}\left(\frac{\partial^2 w}{\partial x\partial \varphi} - \frac{\partial v}{\partial x}\right)$$

In the previous relations, the term R denotes the radius of the shell. In what follows, the previous strain-displacement relations must be implemented to derive the motion equations of shell-type elements. To this purpose, the dynamic form of the principle of virtual work, also known as Hamilton's principle, will be utilized. Implementation of this principle will be observed in all of the following sections dealing with the derivation of the motion equations for various types of continuous systems. This principle states that in any desired time interval, the variation of the total energy of the system must be equal to zero. The total energy is identical with an algebraic summation of the strain energy of the system, its kinetic energy, and work done on the system by external loading. This principle can be formulated in the following form:

$$\delta \int_{t_1}^{t_2} (U - T + V)\, dt = 0 \qquad (2.21)$$

where U, T, and V are strain energy, kinetic energy, and work done by external loading, respectively. Now, each of the abovementioned expressions (i.e., variations of the introduced functions) must be calculated accordingly. All of these mathematical operations must be integrated over the total volume of the element. This integration will be divided into three independent integrals to cover longitudinal, circumferential, and through the thickness directions in the case of investigating a cylindrical shell. Based on the primary definition of the strain energy density (i.e., the double contraction of stress and strain tensors in a desired control volume) and integrating from its variation for linear elastic solids over the total volume, the variation of strain energy can be formulated as follows:

$$\delta U = \int_{-\frac{h}{2}}^{\frac{h}{2}} \int_0^{2\pi} \int_0^L \sigma_{ij}\delta\varepsilon_{ij}\, R\, dx\, d\varphi\, dz \qquad (2.22)$$

The variation of the work done by external loading must be formulated in the following step. Herein, the general form, in which the cylinder is presumed to be loaded in radial, longitudinal, and circumferential directions and rested on a two-parameter Winkler-Pasternak medium, will be considered. The radial, circumferential, and axial loads are shown with N_r, N_φ, and N_x, respectively. In addition, the stiffness coefficients of the springs of elastic medium will be shown with k_w for the Winkler coefficient and k_p for the Pasternak coefficient. Finally, the variation of work done by external loading will be in the following form:

$$\delta V = \int_{-\frac{h}{2}}^{\frac{h}{2}} \int_0^{2\pi} \int_0^L \left(N_r + N_x \frac{\partial^2 w}{\partial x^2} + \frac{N_\varphi}{R^2} \frac{\partial^2 w}{\partial \varphi^2} - k_w w \right.$$
$$\left. + k_p \left(\frac{\partial^2 w}{\partial x^2} + \frac{1}{R^2} \frac{\partial^2 w}{\partial \varphi^2} \right) \right) \delta w R dx d\varphi dz \tag{2.23}$$

In addition, the variation of kinetic energy of the shell can be calculated in the following form:

$$\delta T = \int_{-\frac{h}{2}}^{\frac{h}{2}} \int_0^{2\pi} \int_0^L \rho(z) \left[\frac{\partial u_x}{\partial t} \frac{\partial \delta u_x}{\partial t} + \frac{\partial u_\varphi}{\partial t} \frac{\partial \delta u_\varphi}{\partial t} + \frac{\partial u_z}{\partial t} \frac{\partial \delta u_z}{\partial t} \right] R dx d\varphi dz \tag{2.24}$$

In the equation above, the dependency of the effective mass density on the thickness direction is considered by assuming this term to be a function of the independent variable, z. Now, the Euler-Lagrange equations of motion for the fluid conveying cylindrical shells rested on elastic medium and subjected to various types of radial, axial, and circumferential excitations can be achieved. To this purpose, the variations calculated in Eqs. (2.22), (2.23), and (2.24) must be inserted in the definition of Hamilton's principle (i.e., Eq. (2.21)). This substitution results in an equation that contains three parts due to existence of three degrees of freedom in the kinematic shell theory which is utilized in this study. Each of the parts consists of an expression multiplied by a variation of one of the components of the displacement field of the shell. Due to the fact that the right-hand side of this equation must be equal with zero (remember the definition of the Hamilton's principle), the left-hand side of it must be zero, too. The trivial response of this equation is that we assume the variations of displacement fields to be identical with zero. However, the nontrivial response can be reached setting the expressions behind each of the aforementioned variations to be zero. Hence, the Euler-Lagrange equations can be written in the following form:

$$R \frac{\partial N_{xx}}{\partial x} + \frac{\partial N_{x\varphi}}{\partial \varphi} = I_0 R \frac{\partial^2 u}{\partial t^2} \tag{2.25}$$

$$\frac{1}{R} \frac{\partial M_{\varphi\varphi}}{\partial \varphi} + 2 \frac{\partial M_{x\varphi}}{\partial x} + R \frac{\partial N_{x\varphi}}{\partial x} + \frac{\partial N_{\varphi\varphi}}{\partial \varphi} = I_0 R \frac{\partial^2 v}{\partial t^2} \tag{2.26}$$

$$-N_{\varphi\varphi} + \frac{\partial N_{\varphi\varphi}}{\partial \varphi}\frac{\partial w}{\partial \varphi} + \frac{1}{R}\frac{\partial^2 M_{\varphi\varphi}}{\partial \varphi^2} + N_{\varphi\varphi}\frac{\partial^2 w}{\partial \varphi^2} + \frac{\partial N_{x\varphi}}{\partial x}\frac{\partial w}{\partial \varphi} + \frac{\partial N_{x\varphi}}{\partial \varphi}\frac{\partial w}{\partial x}$$

$$+R\frac{\partial N_{xx}}{\partial x}\frac{\partial w}{\partial x} + 2\frac{\partial^2 M_{x\varphi}}{\partial x \partial \varphi} + 2N_{x\varphi}\frac{\partial^2 w}{\partial x \partial \varphi} + RN_{xx}\frac{\partial^2 w}{\partial x^2} \tag{2.27}$$

$$+R\frac{\partial^2 M_{xx}}{\partial x^2} - k_w Rw + k_p R\left(\frac{\partial^2 w}{\partial x^2} + \frac{1}{R^2}\frac{\partial^2 w}{\partial \varphi^2}\right) = I_0 R\frac{\partial^2 w}{\partial t^2}$$

where stress resultants and mass moments of inertia utilized in the equations above can be calculated using the definition below:

$$\left[N_{xx}, N_{\varphi\varphi}, N_{x\varphi}\right] = \int_{-\frac{h}{2}}^{\frac{h}{2}} \left[\sigma_{xx}, \sigma_{\varphi\varphi}, \sigma_{x\varphi}\right] dz,$$

$$\left[M_{xx}, M_{\varphi\varphi}, M_{x\varphi}\right] = \int_{-\frac{h}{2}}^{\frac{h}{2}} \left[\sigma_{xx}, \sigma_{\varphi\varphi}, \sigma_{x\varphi}\right] z dz \tag{2.28}$$

and

$$I_0 = \int_{-\frac{h}{2}}^{\frac{h}{2}} \rho(z) dz \tag{2.29}$$

To achieve the final governing equations of the problem, the effect of the stress-strain relationship on the stress resultants of the shell must be applied. For this reason, some integrals, the same as those introduced in Eq. (2.28), must be calculated over the thickness. The aforementioned integrals can be expressed by:

$$\begin{bmatrix} N_{xx} \\ N_{\varphi\varphi} \\ N_{x\varphi} \\ M_{xx} \\ M_{\varphi\varphi} \\ M_{x\varphi} \end{bmatrix} = \begin{bmatrix} A_{11} & A_{12} & 0 & B_{11} & B_{12} & 0 \\ A_{12} & A_{22} & 0 & B_{12} & B_{22} & 0 \\ 0 & 0 & A_{66} & 0 & 0 & B_{66} \\ B_{11} & B_{12} & 0 & D_{11} & D_{12} & 0 \\ B_{12} & B_{22} & 0 & D_{12} & D_{22} & 0 \\ 0 & 0 & B_{66} & 0 & 0 & D_{66} \end{bmatrix} \begin{bmatrix} \dfrac{\partial u}{\partial x} + \dfrac{1}{2}\left(\dfrac{\partial w}{\partial x}\right)^2 \\ \dfrac{1}{R}\left(\dfrac{\partial v}{\partial \varphi} + w\right) + \dfrac{1}{2R}\left(\dfrac{\partial w}{\partial \varphi}\right)^2 \\ \dfrac{1}{R}\dfrac{\partial u}{\partial \varphi} + \dfrac{\partial v}{\partial x} + \dfrac{1}{R}\dfrac{\partial w}{\partial x}\dfrac{\partial w}{\partial \varphi} \\ -\dfrac{\partial^2 w}{\partial x^2} \\ -\dfrac{1}{R^2}\left(\dfrac{\partial^2 w}{\partial \varphi^2} - \dfrac{\partial v}{\partial \varphi}\right) \\ -\dfrac{1}{R}\left(\dfrac{\partial^2 w}{\partial x \partial \varphi} - \dfrac{\partial v}{\partial x}\right) \end{bmatrix} \tag{2.30}$$

In the equation above, the cross-sectional rigidities can be determined using the following formulas:

$$[A_{11}, B_{11}, D_{11}] = \int_{-\frac{h}{2}}^{\frac{h}{2}} [1, z, z^2] Q_{11} dz,$$

$$[A_{12}, B_{12}, D_{12}] = \int_{-\frac{h}{2}}^{\frac{h}{2}} [1, z, z^2] Q_{12} dz, \quad (2.31)$$

$$[A_{66}, B_{66}, D_{66}] = \int_{-\frac{h}{2}}^{\frac{h}{2}} [1, z, z^2] Q_{66} dz$$

where Q_{ij} arrays stand for the components of the elasticity tensor of the material implemented in the analysis. These arrays will be introduced later in the chapter (see Section 2.3). Now, the PDEs governing the physics of the problem can be attained substituting for the stress resultants from Eq. (2.30) in Eqs. (2.25)–(2.27). Thus, the governing equations can be achieved in the following form:

$$A_{11} R \left(\frac{\partial^2 u}{\partial x^2} + \frac{\partial w}{\partial x} \frac{\partial^2 w}{\partial x^2} \right) + A_{12} \left(\frac{\partial^2 v}{\partial \varphi \partial x} + \frac{\partial w}{\partial x} + \frac{\partial w}{\partial \varphi} \frac{\partial^2 w}{\partial \varphi \partial x} \right) - B_{11} R \frac{\partial^3 w}{\partial x^3}$$

$$- \frac{B_{12}}{R} \left(\frac{\partial^3 w}{\partial \varphi^2 \partial x} - \frac{\partial^2 v}{\partial \varphi \partial x} \right) + A_{66} \left(\frac{1}{R} \frac{\partial^2 u}{\partial \varphi^2} + \frac{\partial^2 v}{\partial x \partial \varphi} + \frac{1}{R} \left[\frac{\partial^2 w}{\partial x \partial \varphi} \frac{\partial w}{\partial \varphi} + \frac{\partial w}{\partial x} \frac{\partial^2 w}{\partial \varphi^2} \right] \right) \quad (2.32)$$

$$- \frac{B_{66}}{R} \left(\frac{\partial^3 w}{\partial x \partial \varphi^2} - \frac{\partial^2 v}{\partial x \partial \varphi} \right) = I_0 R \frac{\partial^2 u}{\partial t^2}$$

$$\frac{B_{12}}{R} \left(\frac{\partial^2 u}{\partial x \partial \varphi} + \frac{\partial w}{\partial x} \frac{\partial^2 w}{\partial x \partial \varphi} \right) + \frac{B_{12}}{R^2} \left(\frac{\partial^2 v}{\partial \varphi^2} + \frac{\partial w}{\partial \varphi} + \frac{\partial w}{\partial \varphi} \frac{\partial^2 w}{\partial \varphi^2} \right) - \frac{D_{12}}{R} \frac{\partial^3 u}{\partial x^2 \partial \varphi}$$

$$- \frac{D_{22}}{R^3} \left(\frac{\partial^3 w}{\partial \varphi^3} - \frac{\partial^2 v}{\partial \varphi^2} \right) + 2 B_{66} \left(\frac{1}{R} \frac{\partial^2 u}{\partial \varphi \partial x} + \frac{\partial^2 v}{\partial x^2} + \frac{1}{R} \left[\frac{\partial^2 w}{\partial x^2} \frac{\partial w}{\partial \varphi} + \frac{\partial w}{\partial x} \frac{\partial^2 w}{\partial \varphi \partial x} \right] \right)$$

$$- \frac{2 D_{66}}{R} \left(\frac{\partial^3 w}{\partial x^2 \partial \varphi} - \frac{\partial^2 v}{\partial x^2} \right) + A_{66} \left(\frac{\partial^2 u}{\partial \varphi \partial x} + R \frac{\partial^2 v}{\partial x^2} + \left[\frac{\partial^2 w}{\partial x^2} \frac{\partial w}{\partial \varphi} + \frac{\partial w}{\partial x} \frac{\partial^2 w}{\partial \varphi \partial x} \right] \right) \quad (2.33)$$

$$- B_{66} \left(\frac{\partial^3 w}{\partial x^2 \partial \varphi} - \frac{\partial^2 v}{\partial x^2} \right) + A_{12} \left(\frac{\partial^2 u}{\partial x \partial \varphi} + \frac{\partial w}{\partial x} \frac{\partial^2 w}{\partial x \partial \varphi} \right) + \frac{A_{22}}{R} \left(\frac{\partial^2 v}{\partial \varphi^2} + \frac{\partial w}{\partial \varphi} + \frac{\partial w}{\partial \varphi} \frac{\partial^2 w}{\partial \varphi^2} \right)$$

$$- B_{12} \frac{\partial^3 w}{\partial x^2 \partial \varphi} - \frac{B_{22}}{R^2} \left(\frac{\partial^3 w}{\partial \varphi^3} - \frac{\partial^2 v}{\partial \varphi^2} \right) = I_0 R \frac{\partial^2 v}{\partial t^2}$$

$$-A_{12}\left(\frac{\partial u}{\partial x}+\frac{1}{2}\left(\frac{\partial w}{\partial x}\right)^2\right)-\frac{A_{22}}{R}\left(\frac{\partial v}{\partial \varphi}+w+\frac{1}{2}\left(\frac{\partial w}{\partial \varphi}\right)^2\right)+B_{12}\frac{\partial^2 w}{\partial x^2}+\frac{B_{22}}{R^2}\left(\frac{\partial^2 w}{\partial \varphi^2}-\frac{\partial v}{\partial \varphi}\right)$$

$$+A_{12}\left(\frac{\partial^2 u}{\partial x \partial \varphi}\frac{\partial w}{\partial \varphi}+\frac{\partial w}{\partial x}\frac{\partial^2 w}{\partial x \partial \varphi}\frac{\partial w}{\partial \varphi}\right)+\frac{A_{22}}{R}\left(\frac{\partial^2 v}{\partial \varphi^2}\frac{\partial w}{\partial \varphi}+\frac{\partial w}{\partial \varphi}\frac{\partial w}{\partial \varphi}+\frac{\partial w}{\partial \varphi}\frac{\partial^2 w}{\partial \varphi^2}\frac{\partial w}{\partial \varphi}\right)-B_{12}\frac{\partial^3 w}{\partial x^2 \partial \varphi}\frac{\partial w}{\partial \varphi}$$

$$-\frac{B_{12}}{R^2}\left(\frac{\partial^3 w}{\partial \varphi^3}\frac{\partial w}{\partial \varphi}-\frac{\partial^2 v}{\partial \varphi^2}\frac{\partial w}{\partial \varphi}\right)+\frac{B_{12}}{R}\left(\frac{\partial^3 u}{\partial x \partial \varphi^2}+\frac{\partial^2 w}{\partial x \partial \varphi}\frac{\partial^2 w}{\partial x \partial \varphi}+\frac{\partial w}{\partial x}\frac{\partial^3 w}{\partial x \partial \varphi^2}\right)$$

$$+\frac{B_{22}}{R^2}\left(\frac{\partial^3 v}{\partial \varphi^3}+\frac{\partial^2 w}{\partial \varphi^2}+\frac{\partial^2 w}{\partial \varphi^2}\frac{\partial^2 w}{\partial \varphi^2}+\frac{\partial w}{\partial \varphi}\frac{\partial^3 w}{\partial \varphi^3}\right)-\frac{D_{12}}{R}\frac{\partial^4 w}{\partial x^2 \partial \varphi^2}-\frac{D_{22}}{R^3}\left(\frac{\partial^4 w}{\partial \varphi^4}-\frac{\partial^3 v}{\partial \varphi^3}\right)$$

$$+A_{12}\left(\frac{\partial u}{\partial x}\frac{\partial^2 w}{\partial \varphi^2}+\frac{1}{2}\left(\frac{\partial w}{\partial x}\right)^2\frac{\partial^2 w}{\partial \varphi^2}\right)+\frac{A_{22}}{R}\left(\frac{\partial v}{\partial \varphi}\frac{\partial^2 w}{\partial \varphi^2}+w\frac{\partial^2 w}{\partial \varphi^2}+\frac{1}{2}\left(\frac{\partial w}{\partial \varphi}\right)^2\frac{\partial^2 w}{\partial \varphi^2}\right)$$

$$-B_{12}\frac{\partial^2 w}{\partial x^2}\frac{\partial^2 w}{\partial \varphi^2}-\frac{B_{22}}{R^2}\left(\frac{\partial^2 w}{\partial \varphi^2}\frac{\partial^2 w}{\partial \varphi^2}-\frac{\partial v}{\partial \varphi}\frac{\partial^2 w}{\partial \varphi^2}\right)+A_{66}\left(\frac{1}{R}\frac{\partial^2 u}{\partial \varphi \partial x}\frac{\partial w}{\partial \varphi}+\frac{\partial^2 v}{\partial x^2}\frac{\partial w}{\partial \varphi}\right.$$

$$+\frac{1}{R}\left[\frac{\partial^2 w}{\partial x^2}\frac{\partial w}{\partial \varphi}\frac{\partial w}{\partial \varphi}+\frac{\partial w}{\partial x}\frac{\partial^2 w}{\partial \varphi \partial x}\frac{\partial w}{\partial \varphi}\right]\right)-\frac{B_{66}}{R}\left(\frac{\partial^3 w}{\partial x^2 \partial \varphi}\frac{\partial w}{\partial \varphi}-\frac{\partial^2 v}{\partial x^2}\frac{\partial w}{\partial \varphi}\right)$$

$$+A_{66}\left(\frac{1}{R}\frac{\partial^2 u}{\partial \varphi^2}\frac{\partial w}{\partial x}+\frac{\partial^2 v}{\partial x \partial \varphi}\frac{\partial w}{\partial x}+\frac{1}{R}\left[\frac{\partial^2 w}{\partial x \partial \varphi}\frac{\partial w}{\partial \varphi}\frac{\partial w}{\partial x}+\frac{\partial w}{\partial x}\frac{\partial^2 w}{\partial \varphi^2}\frac{\partial w}{\partial x}\right]\right)$$

$$-\frac{B_{66}}{R}\left(\frac{\partial^3 w}{\partial x \partial \varphi^2}\frac{\partial w}{\partial x}-\frac{\partial^2 v}{\partial x \partial \varphi}\frac{\partial w}{\partial x}\right)+A_{11}R\left(\frac{\partial^2 u}{\partial x^2}\frac{\partial w}{\partial x}+\frac{\partial w}{\partial x}\frac{\partial^2 w}{\partial x^2}\frac{\partial w}{\partial x}\right) \qquad (2.34)$$

$$+A_{12}\left(\frac{\partial^2 v}{\partial \varphi \partial x}\frac{\partial w}{\partial x}+\frac{\partial w}{\partial x}\frac{\partial w}{\partial x}+\frac{\partial w}{\partial \varphi}\frac{\partial^2 w}{\partial \varphi \partial x}\frac{\partial w}{\partial x}\right)-B_{11}R\frac{\partial^3 w}{\partial x^3}\frac{\partial w}{\partial x}-\frac{B_{12}}{R}\left(\frac{\partial^3 w}{\partial \varphi^2 \partial x}\frac{\partial w}{\partial x}-\frac{\partial^2 v}{\partial \varphi \partial x}\frac{\partial w}{\partial x}\right)$$

$$+2B_{66}\left(\frac{1}{R}\frac{\partial^3 u}{\partial \varphi^2 \partial x}+\frac{\partial^3 v}{\partial x^2 \partial \varphi}+\frac{1}{R}\left[\frac{\partial^3 w}{\partial x^2 \partial \varphi}\frac{\partial w}{\partial \varphi}+\frac{\partial^2 w}{\partial x^2}\frac{\partial^2 w}{\partial \varphi^2}\right.\right.$$

$$+\left.\left.\frac{\partial^2 w}{\partial x \partial \varphi}\frac{\partial^2 w}{\partial \varphi \partial x}+\frac{\partial w}{\partial x}\frac{\partial^3 w}{\partial \varphi^2 \partial x}\right)\right]-\frac{2D_{66}}{R}\left(\frac{\partial^4 w}{\partial x^2 \partial \varphi^2}-\frac{\partial^3 v}{\partial x^2 \partial \varphi}\right)$$

$$+2A_{66}\left(\frac{1}{R}\frac{\partial u}{\partial \varphi}\frac{\partial^2 w}{\partial x \partial \varphi}+\frac{\partial v}{\partial x}\frac{\partial^2 w}{\partial x \partial \varphi}+\frac{1}{R}\frac{\partial w}{\partial x}\frac{\partial w}{\partial \varphi}\frac{\partial^2 w}{\partial x \partial \varphi}\right)$$

$$-\frac{2B_{66}}{R}\left(\frac{\partial^2 w}{\partial x \partial \varphi}\frac{\partial^2 w}{\partial x \partial \varphi}-\frac{\partial v}{\partial x}\frac{\partial^2 w}{\partial x \partial \varphi}\right)+A_{11}R\left(\frac{\partial u}{\partial x}\frac{\partial^2 w}{\partial x^2}+\frac{1}{2}\left(\frac{\partial w}{\partial x}\right)^2\frac{\partial^2 w}{\partial x^2}\right)+$$

$$A_{12}\left(\frac{\partial v}{\partial \varphi}\frac{\partial^2 w}{\partial x^2}+w\frac{\partial^2 w}{\partial x^2}+\frac{1}{2}\left(\frac{\partial w}{\partial \varphi}\right)^2\frac{\partial^2 w}{\partial x^2}\right)-B_{11}R\frac{\partial^2 w}{\partial x^2}\frac{\partial^2 w}{\partial x^2}-\frac{B_{12}}{R}\left(\frac{\partial^2 w}{\partial \varphi^2}\frac{\partial^2 w}{\partial x^2}-\frac{\partial v}{\partial \varphi}\frac{\partial^2 w}{\partial x^2}\right)$$

$$+B_{11}R\left(\frac{\partial^3 u}{\partial x^3}+\frac{\partial^2 w}{\partial x^2}\frac{\partial^2 w}{\partial x^2}+\frac{\partial w}{\partial x}\frac{\partial^3 w}{\partial x^3}\right)+B_{12}\left(\frac{\partial^3 v}{\partial \varphi \partial x^2}+\frac{\partial^2 w}{\partial x^2}+\frac{\partial^2 w}{\partial \varphi \partial x}\frac{\partial^2 w}{\partial \varphi \partial x}+\frac{\partial w}{\partial \varphi}\frac{\partial^3 w}{\partial \varphi \partial x^2}\right)$$

$$-D_{11}R\frac{\partial^4 w}{\partial x^4}-\frac{D_{12}}{R}\left(\frac{\partial^4 w}{\partial \varphi^2 \partial x^2}-\frac{\partial^3 v}{\partial \varphi \partial x^2}\right)-k_w Rw+k_p R\left(\frac{\partial^2 w}{\partial x^2}+\frac{1}{R^2}\frac{\partial^2 w}{\partial \varphi^2}\right)=I_0 R\frac{\partial^2 w}{\partial t^2}$$

Now, the set of the governing equations must be solved to determine the dynamic response of the system. One can realize that the aforementioned equations are a set of algebraic nonlinear PDEs and they cannot be solved via well-known analytical methods. Hence, the approximation of the dynamic response of the problem must be explored. Due to the time-dependency of the governing equations, one of the most famous perturbation-based methods will be utilized here to reach the frequency response of the system. The method to be implemented is the multiple scales method. Before starting the solution in the time domain, the spatial functions must be inserted in the governing equations to transfer the governing equations in the time domain. Considering the assumption of having simply supported edges at both ends of the shell-type element, the solution to this dynamic problem can be expressed as follows:

$$u(x,\varphi,t) = \sum_{m=1}^{\infty}\sum_{n=1}^{\infty} u_{mn}(t)\cos\left(\frac{m\pi x}{L}\right)\cos(n\varphi),$$

$$v(x,\varphi,t) = \sum_{m=1}^{\infty}\sum_{n=1}^{\infty} v_{mn}(t)\sin\left(\frac{m\pi x}{L}\right)\sin(n\varphi), \qquad (2.35)$$

$$w(x,\varphi,t) = \sum_{m=1}^{\infty}\sum_{n=1}^{\infty} w_{mn}(t)\sin\left(\frac{m\pi x}{L}\right)\cos(n\varphi)$$

Substituting for the displacements from the equation above in Eqs. (2.32)–(2.34) results in the following equations for the structure under observation:

$$C_{11}u_{mn}(t) + C_{12}v_{mn}(t) + C_{13}w_{mn}(t) + C_{14}w_{mn}^{2}(t) = \ddot{u}_{mn}(t) \qquad (2.36)$$

$$C_{21}u_{mn}(t) + C_{22}v_{mn}(t) + C_{23}w_{mn}(t) + C_{24}w_{mn}^{2}(t) = \ddot{v}_{mn}(t) \qquad (2.37)$$

$$C_{31}u_{mn}(t) + C_{32}v_{mn}(t) + C_{33}w_{mn}(t) + C_{34}w_{mn}^{2}(t) + C_{35}w_{mn}^{3}(t)$$
$$+C_{36}u_{mn}(t)w_{mn}(t) + C_{37}v_{mn}(t)w_{mn}(t) = \ddot{w}_{mn}(t) \qquad (2.38)$$

in which the coefficients C_{ij} are known scalars and will be introduced below. It is worth mentioning that the influences of the inertia terms $\ddot{u}_{mn}$ and $\ddot{v}_{mn}$ are very tiny in comparison with the effect of the inertia term in Eq. (2.38). Therefore, in the dynamic analyses the inertia terms will be assumed to be negligible. Following this assumption and considering Eqs. (2.36) and (2.37) as two linear algebraic equation with constant coefficients, the amplitudes u_{mn} and v_{mn} can be derived in the following form:

$$u_{mn}(t) = \frac{C_{23}C_{12} - C_{22}C_{13}}{C_{11}C_{22} - C_{12}C_{21}} w_{mn}(t) + \frac{C_{24}C_{21} - C_{22}C_{14}}{C_{11}C_{22} - C_{12}C_{21}} w_{mn}^{2}(t) \qquad (2.39)$$

$$v_{mn}(t) = \frac{C_{11}[C_{22}C_{13} - C_{23}C_{12}] - C_{13}[C_{11}C_{22} - C_{12}C_{21}]}{C_{12}[C_{11}C_{22} - C_{12}C_{21}]} w_{mn}(t)$$
$$+ \frac{C_{11}[C_{22}C_{14} - C_{24}C_{21}] - C_{14}[C_{11}C_{22} - C_{12}C_{21}]}{C_{12}[C_{11}C_{22} - C_{12}C_{21}]} w_{mn}^{2}(t) \qquad (2.40)$$

The previous equations can be inserted in Eq. (2.38) to reach the following form:

$$\ddot{w}_{mn}(t)+c_1 w_{mn}(t)+c_2 w_{mn}^2(t)+c_3 w_{mn}^3(t)=0 \tag{2.41}$$

The vibration problem of the shells can be solved once the response to the equation above is attained. The initial conditions of the previous equation can be presented in the following form:

$$w_{mn}(0)=w_{max}, \quad \dot{w}_{mn}(0)=0 \tag{2.42}$$

where w_{max} stands for the maximum amount of the function $w_{mn}(t)$. Now the solution procedure can be started. To this purpose, the multiple scales method will be utilized as mentioned before. At first, it is better to introduce the scaled times as follows:

$$T_\alpha = \varepsilon^\alpha t, \quad \alpha=0,1,2,\dots \tag{2.43}$$

In the equation above, the term ε is a dimensionless tiny parameter. According to this method, the first and second derivatives of any desired function with respect to time can be approximated via the following instructions:

$$\begin{cases} \dfrac{\partial}{\partial t}=D_0+\varepsilon D_1+\varepsilon^2 D_2+\cdots \\[2mm] \dfrac{\partial^2}{\partial t^2}=D_0{}^2+2\varepsilon D_0 D_1+\varepsilon^2\left(D_1{}^2+2D_0 D_2\right)+\cdots \end{cases} \tag{2.44}$$

in which the operator D_α can be defined as follows:

$$D_\alpha(.)=\frac{\partial(.)}{\partial T_\alpha} \tag{2.45}$$

Now, the response to the deflection amplitude for the case under observation can be presented as follows:

$$w_{mn}(t)=\varepsilon w_1(T_0,T_1,T_2)+\varepsilon^2 w_2(T_0,T_1,T_2)+\varepsilon^3 w_3(T_0,T_1,T_2) \tag{2.46}$$

Once Eqs. (2.44) and (2.46) are substituted into Eq. (2.41) and setting the coefficient of each power of the tiny parameter ε to be zero, one can reach the following relations:

$$D_0{}^2 w_1+\omega_0{}^2 w_1=0 \tag{2.47}$$

$$D_0{}^2 w_2+\omega_0{}^2 w_2=-2D_0 D_1 w_1-c_2 w_1{}^2 \tag{2.48}$$

$$D_0{}^2 w_3+\omega_0{}^2 w_3=-2D_0 D_1 w_2-2D_0 D_2 w_1-D_1{}^2 w_1-2c_2 w_1 w_2-c_3 w_1{}^3 \tag{2.49}$$

in which the linear natural frequency of the system is

$$\omega_L = \omega_0 = \sqrt{c_1} \tag{2.50}$$

Based on the fundamental methods of differential equations, the solution of Eq. (2.47) can be presented in the following form:

$$w_1 = A(T_1,T_2)e^{i\omega_0 T_0} + \overline{A}(T_1,T_2)e^{-i\omega_0 T_0} \tag{2.51}$$

where $A(T_1,T_2)$ and $\overline{A}(T_1,T_2)$ are complex conjugates (CC). Once the equation above is inserted into Eq. (2.48), the following equation results:

$$D_0{}^2 w_2 + \omega_0{}^2 w_2 = -2i\omega_0 D_1 A e^{i\omega_0 T_0} - c_2 A^2 e^{2i\omega_0 T_0} - c_2 A\overline{A} + CC \tag{2.52}$$

in which CC denotes the complex conjugate of the previous terms. It is obvious that the secular term of the above equation (i.e., $-2i\omega_0 D_1 A e^{i\omega_0 T_0}$) must be eliminated; so, the following relation is valid:

$$D_1 A = 0 \tag{2.53}$$

Now the solution of Eq. (2.52) can be expressed in the following format:

$$w_2 = \frac{c_2}{3\omega_0{}^2} A^2 e^{2i\omega_0 T_0} - \frac{c_2}{\omega_0{}^2} A\overline{A} + CC \tag{2.54}$$

The solutions of w_1 and w_2 can be substituted from Eqs. (2.51) and (2.54) into Eq. (2.49) to reach the following equation:

$$D_0{}^2 w_3 + \omega_0{}^2 w_3 = -\left[c_3 A^3 + \frac{2c_2{}^2}{3\omega_0{}^2} A^3 \right] e^{3i\omega_0 T_0}$$
$$+ \left[-3c_3 A^2 \overline{A} + \frac{10c_2{}^2}{3\omega_0{}^2} A^2 \overline{A} - 2i\omega_0 D_2 A \right] e^{i\omega_0 T_0} + CC \tag{2.55}$$

In the equation above, the secular terms must be eliminated. Hence, we can write the following relation:

$$-3c_3 A^2 \overline{A} + \frac{10c_2{}^2}{3\omega_0{}^2} A^2 \overline{A} - 2i\omega_0 D_2 A = 0 \tag{2.56}$$

The solution of the equation above can be defined in the following form:

$$A = \frac{1}{2} a e^{i\phi} \tag{2.57}$$

Once the recommended solution, presented in Eq. (2.57), is inserted into Eq. (2.56) and the real and imaginary parts of the enriched equation are separated, one can reach the following set of algebraic equations:

$$\begin{cases} \omega_0 \dfrac{\partial a}{\partial T_2} = 0 \\[4mm] \dfrac{5c_2^2}{12\omega_0^2} a^3 - \dfrac{3c_3}{8} a^3 + \omega_0 a \dfrac{\partial \phi}{\partial T_2} = 0 \end{cases} \tag{2.58}$$

Solving the equation above results in the following equation for the variant ϕ:

$$\phi = \frac{9c_3\omega_0^2 - 10c_2^2}{24\omega_0^3} a^2 T_2 + \phi_0 \tag{2.59}$$

In the previous equation, ϕ_0 is a constant phase angle. Substituting for ϕ from Eq. (2.59) in Eq. (2.57) and regarding the primary definition of the scaled time T_2 (see Eq. (2.43)), the following formula can be attained:

$$A = \frac{1}{2} a \exp\left[i \frac{9c_3\omega_0^2 - 10c_2^2}{24\omega_0^3} a^2 \varepsilon^2 t + i\phi_0 \right] \tag{2.60}$$

Inserting Eqs. (2.51), (2.54), and (2.60) in Eq. (2.46) reveals:

$$w_{mn}(t) = \varepsilon a \cos\left(\omega_{NL} t + \phi_0\right) - \frac{\varepsilon^2 a^2 c_2}{2c_1}\left[1 - \frac{1}{3}\cos\left(2\omega_{NL} t + 2\phi_0\right)\right] + O(\varepsilon^3) \tag{2.61}$$

in which the coefficients c_i ($i = 1,2,3$) can be defined in the following form:

$$c_1 = \frac{C_{21}\left[C_{32}C_{13} - C_{12}C_{33}\right]}{\left[C_{12}C_{21} - C_{11}C_{22}\right]} + \frac{C_{22}\left[C_{11}C_{33} - C_{13}C_{31}\right]}{\left[C_{12}C_{21} - C_{11}C_{22}\right]} + \frac{C_{23}\left[C_{12}C_{31} - C_{11}C_{32}\right]}{\left[C_{12}C_{21} - C_{11}C_{22}\right]}, \tag{2.62}$$

$$c_2 = 0,$$

$$c_3 = C_{35}$$

Once the initial conditions, defined in Eq. (2.42), are applied, the following relations can be achieved:

$$\phi_0 = 0, \quad a\varepsilon = w_{max} \tag{2.63}$$

In the end, the nonlinear natural frequency of the shell-type element can be depicted in the following form:

$$\omega_{NL} = \omega_L\left[1 + \frac{9c_3 c_1 - 10c_2^2}{24c_1^2} w_{max}^2\right] + O(\varepsilon^3) \tag{2.64}$$

2.2.2 FIRST-ORDER SHEAR DEFORMATION SHELL THEORY

Based on this theory, the influences of shear stress and strain will be approximated
using a constant distribution for these variables across the thickness direction.
Indeed, a shear correction coefficient will be employed to estimate the distribution of shear stress and strain once the thickness of the shell varies. This theory is
better than the classical theory of cylindrical shells due to its potential to consider
shear deformation up to first-order, whereas such an effect is not included in the
classical theory. The displacement field for a cylindrical shell can be expressed
as follows:

$$u_x(x,\varphi,z,t) = u(x,\varphi,t) + z\theta_x(x,\varphi,t),$$
$$u_\varphi(x,\varphi,z,t) = v(x,\varphi,t) + z\theta_\varphi(x,\varphi,t), \tag{2.65}$$
$$u_z(x,\varphi,z,t) = w(x,\varphi,t)$$

where u denotes the component of displacement parallel with the axis of the shell
and v stands for the circumferential displacement. Also, w is the lateral deflection
of the shell. On the other hand, θ_x and θ_φ are the rotation components about axial
and circumferential directions, respectively. The strain-displacement relations of the
cylindrical shell can be obtained using the preliminary definitions explained in
the Section 2.1. In fact, using Eq. (2.18) and considering the displacement field of the
shell as the same as that introduced in Eq. (2.65), the nonzero arrays of strain tensor
can be achieved in the following form:

$$\varepsilon_{xx} = \frac{1}{2}\left(\frac{\partial u_x}{\partial x} + \frac{\partial u_x}{\partial x}\right) = \frac{\partial u}{\partial x} + z\frac{\partial \theta_x}{\partial x},$$

$$\varepsilon_{\varphi\varphi} = \frac{1}{2}\left(\frac{\partial u_\varphi}{\partial \varphi} + \frac{\partial u_\varphi}{\partial \varphi}\right) = \frac{1}{R}\left(\frac{\partial v}{\partial \varphi} + z\frac{\partial \theta_\varphi}{\partial \varphi} + w\right),$$

$$\varepsilon_{x\varphi} = \frac{1}{2}\left(\frac{\partial u_x}{\partial \varphi} + \frac{\partial u_\varphi}{\partial x}\right) = \frac{1}{R}\frac{\partial u}{\partial \varphi} + \frac{\partial v}{\partial x} + \frac{z}{R}\frac{\partial \theta_x}{\partial \varphi} + z\frac{\partial \theta_\varphi}{\partial x}, \tag{2.66}$$

$$\varepsilon_{xz} = \frac{1}{2}\left(\frac{\partial u_x}{\partial z} + \frac{\partial u_z}{\partial x}\right) = \theta_x + \frac{\partial w}{\partial x},$$

$$\varepsilon_{\varphi z} = \frac{1}{2}\left(\frac{\partial u_\varphi}{\partial z} + \frac{\partial u_z}{\partial \varphi}\right) = \theta_\varphi + \frac{1}{R}\frac{\partial w}{\partial \varphi} - \frac{v}{R}$$

in which R is the radius of the cylinder and the gradient operation is completed in the
polar coordinate system. As explained in the previous subsection (see Section 2.2.1),
the motion equations will be derived for the case of a time-dependent problem on the
basis of Hamilton's principle. Following the procedure as before, the variation of
the strain energy of the shell based on the first-order shear deformation theory can

be expressed as the same as Eq. (2.22). However, the effects of shear stress and strain will be included here.

The variation of work done by external loading must be calculated. As expressed in Eq. (2.23), the influences of different radial, longitudinal, and circumferential loadings will be considered in this section as well as those of the elastic foundation. However, in this section the effect of the viscose fluid flowing inside the cylinder will be considered, too. To this purpose, it is necessary to introduce the involved parameters before showing the effect of the viscose flow on the variation of the work done by external loading on the structure. Indeed, the density of the conveying fluid, the flow velocity, and viscosity of the fluid will be shown with ρ_b, V_R, and μ, respectively. The fluid flow will be considered to be symmetric, laminar, and Newtonian. Therefore, the impact of this flow on the motion equations of the cylinder must be applied in the framework of extending the Navier-Stokes equation for the flow under observation. Finally, the variation of work done by external loading will be in the following form:

$$\delta V = \int_{-\frac{h}{2}}^{\frac{h}{2}} \int_0^{2\pi} \int_0^L \left(\begin{array}{c} N_r + N_x \dfrac{\partial^2 w}{\partial x^2} + \dfrac{N_\varphi}{R^2} \dfrac{\partial^2 w}{\partial \varphi^2} - k_w w \\[3mm] + k_p \left(\dfrac{\partial^2 w}{\partial x^2} + \dfrac{1}{R^2} \dfrac{\partial^2 w}{\partial \varphi^2} \right) \\[3mm] + \dfrac{\mu}{R^3} \dfrac{\partial^2 V_R}{\partial \varphi^2} - \dfrac{2\mu}{R^2} V_R + \mu \dfrac{\partial^2 V_R}{\partial x^2} - \rho_h \dfrac{d^2 w}{dt^2} \end{array} \right) \delta w R dx d\varphi dz \quad (2.67)$$

Then the variation of the kinetic energy of the system must be computed. This variation was presented in Eq. (2.24) and it is avoided now to present it at another time in this section. Now, the Euler-Lagrange equations of motion for the fluid conveying cylindrical shells rested on elastic medium and subjected to various types of radial, axial, and circumferential excitations can be achieved. To this purpose, the variations calculated in Eqs. (2.22), (2.67), and (2.24) must be inserted into the definition of Hamilton's principle (i.e., Eq. (2.21)). Doing so results in an equation which contains five parts due to the existence of five degrees of freedom in the kinematic shell theory that is utilized in this study. As mentioned before, each of the parts consists of an expression multiplied by the variation of one of the components of the displacement field of the shell. Once the nontrivial responses of the enriched equation are collected, the Euler-Lagrange equations can be written in the following form:

$$\frac{\partial N_{xx}}{\partial x} + \frac{1}{R} \frac{\partial N_{x\varphi}}{\partial \varphi} = I_0 \frac{\partial^2 u}{\partial t^2} + I_1 \frac{\partial^2 \theta_x}{\partial t^2} \quad (2.68)$$

$$\frac{\partial N_{x\varphi}}{\partial x} + \frac{1}{R} \frac{\partial N_{\varphi\varphi}}{\partial \varphi} + \frac{Q_{z\varphi}}{R} = I_0 \frac{\partial^2 v}{\partial t^2} + I_1 \frac{\partial^2 \theta_\varphi}{\partial t^2} \quad (2.69)$$

$$\frac{\partial Q_{xz}}{\partial x} + \frac{1}{R}\frac{\partial Q_{z\varphi}}{\partial \varphi} - \frac{N_{\varphi\varphi}}{R} + N_r + N_x \frac{\partial^2 w}{\partial x^2} + \frac{N_\varphi}{R^2}\frac{\partial^2 w}{\partial \varphi^2} - k_w w + k_p \left(\frac{\partial^2 w}{\partial x^2} + \frac{1}{R^2}\frac{\partial^2 w}{\partial \varphi^2} \right)$$

$$-\rho_b h_f v_x^2 \frac{\partial^2 w}{\partial x^2} + \mu_f h_f v_x \left[\frac{1}{R^2}\left(\frac{\partial^3 w}{\partial x \partial \varphi^2} - 2\frac{\partial w}{\partial x} \right) + \frac{\partial^3 w}{\partial x^3} \right] = I_0 \frac{\partial^2 w}{\partial t^2} \qquad (2.70)$$

$$+\rho_b h_f \left(\frac{\partial^2 w}{\partial t^2} + 2v_x \frac{\partial^2 w}{\partial x \partial t} \right) - \mu_f h_f \left[\frac{1}{R^2}\left(2\frac{\partial w}{\partial t} - \frac{\partial^3 w}{\partial t \partial \varphi^2} \right) - \frac{\partial^3 w}{\partial t \partial x^2} \right]$$

$$\frac{\partial M_{xx}}{\partial x} + \frac{1}{R}\frac{\partial M_{x\varphi}}{\partial \varphi} - Q_{xz} = I_1 \frac{\partial^2 u}{\partial t^2} + I_2 \frac{\partial^2 \theta_x}{\partial t^2} \qquad (2.71)$$

$$\frac{\partial M_{x\varphi}}{\partial x} + \frac{1}{R}\frac{\partial M_{\varphi\varphi}}{\partial \varphi} - Q_{\varphi z} = I_1 \frac{\partial^2 v}{\partial t^2} + I_2 \frac{\partial^2 \theta_\varphi}{\partial t^2} \qquad (2.72)$$

Some of the stress resultants existing in the equations above were previously introduced in Eq. (2.28); however, some of them, related to the impacts of shear stress and strain, are new. Now, all of the stress resultants available in Eqs. (2.68)–(2.72) will be presented together for the sake of completeness:

$$\left[N_{xx}, N_{\varphi\varphi}, N_{x\varphi} \right] = \int_{-\frac{h}{2}}^{\frac{h}{2}} \left[\sigma_{xx}, \sigma_{\varphi\varphi}, \sigma_{x\varphi} \right] dz,$$

$$\left[M_{xx}, M_{\varphi\varphi}, M_{x\varphi} \right] = \int_{-\frac{h}{2}}^{\frac{h}{2}} \left[\sigma_{xx}, \sigma_{\varphi\varphi}, \sigma_{x\varphi} \right] z dz, \qquad (2.73)$$

$$\left[Q_{xz}, Q_{z\varphi} \right] = \kappa_s \int_{-\frac{h}{2}}^{\frac{h}{2}} \left[\sigma_{xz}, \sigma_{z\varphi} \right] dz$$

Also, the mass moments of inertia utilized in the equations above can be expressed in the following form:

$$\left[I_0, I_1, I_2 \right] = \int_{-\frac{h}{2}}^{\frac{h}{2}} \left[1, z, z^2 \right] \rho(z) dz \qquad (2.74)$$

In Eq. (2.73), the term κ_s stands for the shear correction factor of the FSDT which equals 5/6 in general. To achieve the final governing equations of the problem, the effect of the stress-strain relationship on the stress resultants of the shell must be applied. For this reason, some integrals must be calculated over the thickness, the same as those introduced in Eq. (2.73). The aforementioned integrals can be expressed by:

$$
\begin{bmatrix} N_{xx} \\ M_{xx} \\ N_{\varphi\varphi} \\ M_{\varphi\varphi} \end{bmatrix} = \begin{bmatrix} A_{11} & B_{11} & \dfrac{A_{12}}{R} & \dfrac{B_{12}}{R} \\ B_{11} & D_{11} & \dfrac{B_{12}}{R} & \dfrac{D_{12}}{R} \\ A_{12} & B_{12} & \dfrac{A_{11}}{R} & \dfrac{B_{11}}{R} \\ B_{12} & D_{12} & \dfrac{B_{11}}{R} & \dfrac{D_{11}}{R} \end{bmatrix} \begin{bmatrix} \dfrac{\partial u}{\partial x} \\ \dfrac{\partial \theta_x}{\partial x} \\ \dfrac{\partial v}{\partial \varphi} + w \\ \dfrac{\partial \theta_\varphi}{\partial \varphi} \end{bmatrix},
$$

(2.75)

$$
\begin{bmatrix} N_{x\varphi} \\ M_{x\varphi} \end{bmatrix} = \begin{bmatrix} A_{66} & B_{66} \\ B_{66} & D_{66} \end{bmatrix} \begin{bmatrix} \dfrac{1}{R}\dfrac{\partial u}{\partial \varphi} + \dfrac{\partial v}{\partial x} \\ \dfrac{1}{R}\dfrac{\partial \theta_x}{\partial \varphi} + \dfrac{\partial \theta_\varphi}{\partial x} \end{bmatrix},
$$

$$
Q_{xz} = A_{55}^s \left(\theta_x + \frac{\partial w}{\partial x} \right), \quad Q_{\varphi z} = A_{55}^s \left(\theta_\varphi + \frac{1}{R}\frac{\partial w}{\partial \varphi} - \frac{v}{R} \right)
$$

in which the cross-sectional rigidities can be defined in the following form:

$$
[A_{11}, B_{11}, D_{11}] = \int_{-\frac{h}{2}}^{\frac{h}{2}} [1, z, z^2] Q_{11} dz,
$$

$$
[A_{12}, B_{12}, D_{12}] = \int_{-\frac{h}{2}}^{\frac{h}{2}} [1, z, z^2] Q_{12} dz,
$$

(2.76)

$$
[A_{66}, B_{66}, D_{66}] = \int_{-\frac{h}{2}}^{\frac{h}{2}} [1, z, z^2] Q_{66} dz,
$$

$$
A_{55}^s = \kappa_s \int_{-\frac{h}{2}}^{\frac{h}{2}} Q_{55} dz
$$

Most of the rigidities defined above were previously introduced in Eq. (2.31); however, they are represented here for the sake of completeness. As you know from the previous section, the Q_{ij} arrays are the components of the elasticity tensor of the material that the continuum is manufactured from. These quantities will be introduced in the following steps related to the homogenization of the nanocomposite materials. Now the motion equations and the relation between the stress resultants and displacement field of the structure are both in hand. Hence, the governing equations of the problem can be obtained by substituting for the stress resultants from Eq. (2.75) in Eqs. (2.68)–(2.72). Once this substitution is completed, the governing equations of the shell can be presented in the following form:

$$A_{11}\frac{\partial^2 u}{\partial x^2} + B_{11}\frac{\partial^2 \theta_x}{\partial x^2} + \frac{A_{12}}{R}\left(\frac{\partial^2 v}{\partial x \partial \varphi} + \frac{\partial w}{\partial x}\right) + \frac{B_{12}}{R}\frac{\partial^2 \theta_\varphi}{\partial x \partial \varphi} + \frac{A_{66}}{R}\left(\frac{1}{R}\frac{\partial^2 u}{\partial \varphi^2} + \frac{\partial^2 v}{\partial x \partial \varphi}\right)$$

$$+\frac{B_{66}}{R}\left(\frac{1}{R}\frac{\partial^2 \theta_x}{\partial \varphi^2} + \frac{\partial^2 \theta_\varphi}{\partial x \partial \varphi}\right) - I_0\frac{\partial^2 u}{\partial t^2} - I_1\frac{\partial^2 \theta_x}{\partial t^2} = 0 \tag{2.77}$$

$$A_{66}\left(\frac{1}{R}\frac{\partial^2 u}{\partial x \partial \varphi} + \frac{\partial^2 v}{\partial x^2}\right) + B_{66}\left(\frac{1}{R}\frac{\partial^2 \theta_x}{\partial x \partial \varphi} + \frac{\partial^2 \theta_\varphi}{\partial x^2}\right) + \frac{A_{12}}{R}\frac{\partial^2 u}{\partial x \partial \varphi} + \frac{B_{12}}{R}\frac{\partial^2 \theta_x}{\partial x \partial \varphi}$$

$$+\frac{A_{11}}{R^2}\left(\frac{\partial^2 v}{\partial \varphi^2} + \frac{\partial w}{\partial \varphi}\right) + \frac{B_{11}}{R^2}\frac{\partial^2 \theta_\varphi}{\partial \varphi^2} + \frac{A_{55}^s}{R}\left(\theta_\varphi + \frac{1}{R}\frac{\partial w}{\partial \varphi} - \frac{v}{R}\right) - I_0\frac{\partial^2 v}{\partial t^2} - I_1\frac{\partial^2 \theta_\varphi}{\partial t^2} = 0 \tag{2.78}$$

$$A_{55}^s\left(\frac{\partial \theta_x}{\partial x} + \frac{\partial^2 w}{\partial x^2}\right) + \frac{A_{55}^s}{R}\left(\frac{\partial \theta_\varphi}{\partial \varphi} + \frac{1}{R}\frac{\partial^2 w}{\partial \varphi^2} - \frac{1}{R}\frac{\partial v}{\partial \varphi}\right) - \frac{A_{12}}{R}\frac{\partial u}{\partial x} - \frac{B_{12}}{R}\frac{\partial \theta_x}{\partial x}$$

$$-\frac{A_{11}}{R^2}\left(\frac{\partial v}{\partial \varphi} + w\right) - \frac{B_{11}}{R^2}\frac{\partial \theta_\varphi}{\partial \varphi} + N_r + N_x\frac{\partial^2 w}{\partial x^2} + \frac{N_\varphi}{R^2}\frac{\partial^2 w}{\partial \varphi^2}$$

$$-k_w w + k_p\left(\frac{\partial^2 w}{\partial x^2} + \frac{1}{R^2}\frac{\partial^2 w}{\partial \varphi^2}\right) - \rho_b h_f v_x^2\frac{\partial^2 w}{\partial x^2} +$$

$$\mu_f h_f v_x\left[\frac{1}{R^2}\left(\frac{\partial^3 w}{\partial x \partial \varphi^2} - 2\frac{\partial w}{\partial x}\right) + \frac{\partial^3 w}{\partial x^3}\right] - I_0\frac{\partial^2 w}{\partial t^2} - \rho_b h_f\left(\frac{\partial^2 w}{\partial t^2} + 2v_x\frac{\partial^2 w}{\partial x \partial t}\right)$$

$$+\mu_f h_f\left[\frac{1}{R^2}\left(2\frac{\partial w}{\partial t} - \frac{\partial^3 w}{\partial t \partial \varphi^2}\right) - \frac{\partial^3 w}{\partial t \partial x^2}\right] = 0 \tag{2.79}$$

$$B_{11}\frac{\partial^2 u}{\partial x^2} + D_{11}\frac{\partial^2 \theta_x}{\partial x^2} + \frac{B_{12}}{R}\left(\frac{\partial^2 v}{\partial x \partial \varphi} + \frac{\partial w}{\partial x}\right) + \frac{D_{12}}{R}\frac{\partial^2 \theta_\varphi}{\partial x \partial \varphi}$$

$$+\frac{B_{66}}{R}\left(\frac{1}{R}\frac{\partial^2 u}{\partial \varphi^2} + \frac{\partial^2 v}{\partial x \partial \varphi}\right) + \frac{D_{66}}{R}\left(\frac{1}{R}\frac{\partial^2 \theta_x}{\partial \varphi^2} + \frac{\partial^2 \theta_\varphi}{\partial x \partial \varphi}\right) \tag{2.80}$$

$$-A_{55}^s\left(\theta_x + \frac{\partial w}{\partial x}\right) - I_1\frac{\partial^2 u}{\partial t^2} - I_2\frac{\partial^2 \theta_x}{\partial t^2} = 0$$

$$B_{66}\left(\frac{1}{R}\frac{\partial^2 u}{\partial x \partial \varphi}+\frac{\partial^2 v}{\partial x^2}\right)+D_{66}\left(\frac{1}{R}\frac{\partial^2 \theta_x}{\partial x \partial \varphi}+\frac{\partial^2 \theta_\varphi}{\partial x^2}\right)+\frac{B_{12}}{R}\frac{\partial^2 u}{\partial x \partial \varphi}$$

$$+\frac{D_{12}}{R}\frac{\partial^2 \theta_x}{\partial x \partial \varphi}+\frac{B_{11}}{R^2}\left(\frac{\partial^2 v}{\partial \varphi^2}+\frac{\partial w}{\partial \varphi}\right)+\frac{D_{11}}{R^2}\frac{\partial^2 \theta_\varphi}{\partial \varphi^2} \qquad (2.81)$$

$$-A_{55}^s\left(\theta_\varphi+\frac{1}{R}\frac{\partial w}{\partial \varphi}-\frac{v}{R}\right)-I_1\frac{\partial^2 v}{\partial t^2}-I_2\frac{\partial^2 \theta_\varphi}{\partial t^2}=0$$

The problem's formulation is now complete and solving Eqs. (2.77)–(2.81) results in the mechanical response of the problem. Before beginning to discuss about how to solve the set of governing equations, it is necessary to talk about the answers which can be achieved using these equations. Solving the equations above within both spatial and temporal domain results in obtaining the natural frequency of the cylindrical shell under defined loading condition. However, the question remains: How can the buckling load be derived? The answer to this question is not too complex. In fact, the static buckling problem of the governing equations can be solved while all of the differentiations with respect to time are ignored and the axial load applied on the structure is assumed to be identical with N^b, which is the buckling load applied on the shell. Therefore, solving the governing equations for N^b results in determining the critical buckling load of the cylindrical shell. In what follows, an analytical solution will be introduced to solve both static and dynamic problems of shells subjected to such loading conditions. In the case of solving dynamic problem, the displacement fields can be considered to have the following form:

$$u = \sum_{m=1}^{\infty}\sum_{n=1}^{\infty}U_{mn}\frac{\partial X_m(x)}{\partial x}\cos\left(n\varphi\right)e^{i\omega_{mn}t},$$

$$v = \sum_{m=1}^{\infty}\sum_{n=1}^{\infty}V_{mn}X_m(x)\sin\left(n\varphi\right)e^{i\omega_{mn}t},$$

$$w = \sum_{m=1}^{\infty}\sum_{n=1}^{\infty}W_{mn}X_m(x)\cos\left(n\varphi\right)e^{i\omega_{mn}t}, \qquad (2.82)$$

$$\theta_x = \sum_{m=1}^{\infty}\sum_{n=1}^{\infty}\Theta_{xmn}\frac{\partial X_m(x)}{\partial x}\cos\left(n\varphi\right)e^{i\omega_{mn}t},$$

$$\theta_\varphi = \sum_{m=1}^{\infty}\sum_{n=1}^{\infty}\Theta_{\varphi mn}X_m(x)\sin\left(n\varphi\right)e^{i\omega_{mn}t}$$

where m is the longitudinal mode number and n stands for the circumferential wave number. In addition, U_{mn}, V_{mn}, W_{mn}, Θ_{xmn}, and $\Theta_{\varphi mn}$ are the vibration amplitudes. It is noteworthy that $X_m(x)$ is a function that is responsible for satisfying the BCs in the axial direction; also, the natural frequency of the system is ω_{mn}. After inserting Eq. (2.82) in Eqs. (2.77)–(2.81), the following eigenvalue problem can be achieved:

$$\left[K + C\omega_{mn} - M\omega_{mn}^2 \right] \Delta = 0 \tag{2.83}$$

where K, C, and M are stiffness, damping, and mass matrices, respectively and the column vector Δ denotes the amplitude vector. The equation is a 5×5 eigenvalue equation that must be solved to reach the natural frequency of the shell. To solve the equation, the following determinant must be set to zero:

$$\left| K + C\omega_{mn} - M\omega_{mn}^2 \right| = 0 \tag{2.84}$$

Once the previous equation is solved for ω_{mn}, the natural frequency of the vibrating system can be developed. It must be mentioned again that the static stability problem of the formulations presented in Eqs. (2.77)–(2.81) can be solved by setting the determinant of the stiffness matrix to be equal with zero. In fact, in the case of buckling analysis, there is a buckling load in the stiffness matrix and once the determinant of this matrix is set to zero and the equation is solved for that buckling load, the critical buckling load of the shell will be reached. To clarify, the final equation that results in reaching the critical buckling load of the shell will be presented below:

$$\left| K \right| = 0 \tag{2.85}$$

It is worth mentioning that the function $X_m(x)$ must be defined to enable us to derive the mechanical response of either static or dynamic problem. As previously stated, this function must satisfy the BCs at the ends, so it must be able to satisfy the mathematical representation of such conditions. In the case of analyzing a shell with simply supported ends, $X_m(x)$ must be able to satisfy the following equations:

$$w = \frac{\partial^2 w}{\partial x^2} = 0 \quad \text{at} \quad x = 0, L \tag{2.86}$$

Considering the assumptions above and substituting them in Eq. (2.82), the function $X_m(x)$ can be assumed to be in the following form:

$$X_m(x) = sin\left(\frac{m\pi x}{L} \right) \tag{2.87}$$

2.2.3 FIRST-ORDER SHEAR DEFORMATION BEAM THEORY

This section will present a mathematical framework for both static and dynamic problems of nanocomposite beams based on the first-order shear deformation beam hypothesis or the well-known Timoshenko beam theory. Similar to the previous section, which dealt with the mechanical behaviors of shells according to the first-order kinematic shell theory, in this section a constant approximation will be considered for the distribution of shear stress and strain through the thickness direction. Again, a shear correction factor will be introduced and a better estimation will be presented for the mechanical analysis of beam-type elements in comparison with the well-known classical theory of beams, known as Euler-Bernoulli beam theory. The displacement field of the beam within the framework of Timoshenko beam theory can be assumed to be in the following form:

$$u_x(x,z,t) = u(x,t) + z\varphi(x,t),$$
$$u_z(x,z,t) = w(x,t)$$

(2.88)

where u and w are longitudinal displacement and deflection of the beam, respectively. Also, the term φ stands for the total bending rotation of the beam's cross-section. The strain-displacement relations of Timoshenko beams can be achieved using the previous relations. Substituting for the displacement field from Eq. (2.88) in Eq. (2.18), the nonzero arrays of strain tensor for a Timoshenko beam can be reached as follows:

$$\varepsilon_{xx} = \frac{1}{2}\left(\frac{\partial u_x}{\partial x} + \frac{\partial u_x}{\partial x}\right) = \frac{\partial u}{\partial x} + z\frac{\partial^2 \varphi}{\partial x^2},$$

$$\gamma_{xz} = 2\varepsilon_{xz} = \left(\frac{\partial u_x}{\partial z} + \frac{\partial u_z}{\partial x}\right) = \varphi + \frac{\partial w}{\partial x}$$

(2.89)

where ε_{xx} and γ_{xz} are the normal and shear strains, respectively. Following this, the motion equations of such beam-type elements must be obtained using the previously introduced energy-based approach. The dynamic form of the equations will then be extracted and we will explain how to arrive at the motion equations for the case of a static stability problem. Hamilton's principle will be implemented again to derive the motion equations. Based on the Timoshenko beam theory, the variation of the strain energy can be calculated using the following expression:

$$\delta U = \int_\forall \sigma_{ij}\delta\varepsilon_{ij}d\forall = \int_\forall \left(\sigma_{xx}\delta\varepsilon_{xx} + \sigma_{xz}\delta\gamma_{xz}\right)d\forall$$

(2.90)

in which $\forall$ denotes the total volume of the structure. Once the strain-displacement relations introduced in Eq. (2.89) are inserted in the equation above, the following expression can be reached for the variation of the strain energy of the beam:

$$\delta U = \int_0^L \left[N\left(\delta\frac{\partial u}{\partial x}\right) + M\left(\delta\frac{\partial \varphi}{\partial x}\right) + Q\left(\delta\left[\frac{\partial w}{\partial x} + \varphi\right]\right)\right]dx$$

(2.91)

where N, M, and Q are normal force, bending moment, and shear force, respectively. The stress resultants can be defined as follows:

$$N = \int_A \sigma_{xx} dA,$$

$$M = \int_A z\sigma_{xx} dA, \qquad (2.92)$$

$$Q = \kappa_s \int_A \sigma_{xz} dA$$

in which κ_s is the shear correction factor of the FSDT and is equal to 5/6. Now the variation of the kinetic energy of the beam must be formulated. To this purpose, the following expression can be employed:

$$\delta T = \int_\forall \rho(z) \left[\frac{\partial u_x}{\partial t} \frac{\partial \delta u_x}{\partial t} + \frac{\partial u_z}{\partial t} \frac{\partial \delta u_z}{\partial t} \right] d\forall \qquad (2.93)$$

Using the definition of components of displacement field (see Eq. (2.88)), Eq. (2.93) can be rewritten as follows:

$$\delta T = \int_0^L \left[I_0 \left(\frac{\partial u}{\partial t} \frac{\partial \delta u}{\partial t} + \frac{\partial w}{\partial t} \frac{\partial \delta w}{\partial t} \right) + I_1 \left(\frac{\partial \varphi}{\partial t} \frac{\partial \delta u}{\partial t} + \frac{\partial u}{\partial t} \frac{\partial \delta \varphi}{\partial t} \right) + I_2 \frac{\partial \varphi}{\partial t} \frac{\partial \delta \varphi}{\partial t} \right] dx \quad (2.94)$$

in which the mass moments of inertia used in the equation above can be defined as follows:

$$[I_0, I_1, I_2] = \int_A \left[1, z, z^2 \right] \rho(z) dA \qquad (2.95)$$

Now the work done by external loading must be calculated. Assume a Timoshenko beam which is subjected to a thermal loading, N^T. In this case, the variation of work done by this loading can be computed using the following formula:

$$\delta V = \int_0^L N^T \frac{\partial^2 w}{\partial x^2} \delta w dx \qquad (2.96)$$

All of the variations required to construct the motion equations of Timoshenko beams subjected to thermal loading are now in hand. The Euler-Lagrange equations can be easily obtained once Eqs. (2.91), (2.94), and (2.96) are inserted in Eq. (2.21). When this substitution is completely carried out and the nontrivial response of the final equation is selected (see Section 2.2.1 for more discussion about this issue), the Euler-Lagrange equations of Timoshenko beams can be written in the following form:

$$\frac{\partial N}{\partial x} = I_0 \frac{\partial^2 u}{\partial t^2} + I_1 \frac{\partial^2 \varphi}{\partial t^2} \tag{2.97}$$

$$\frac{\partial Q}{\partial x} - N^T \frac{\partial^2 w}{\partial x^2} = I_0 \frac{\partial^2 w}{\partial t^2} \tag{2.98}$$

$$\frac{\partial M}{\partial x} - Q = I_1 \frac{\partial^2 u}{\partial t^2} + I_2 \frac{\partial^2 \varphi}{\partial t^2} \tag{2.99}$$

The impact of the stress-strain relationship of the beam on the stress resultants must be expressed in mathematical language to be able to derive the final governing differential equations. Therefore a group of integrals over the cross-sectional area of the beam must be calculated to reach a representation of the stress resultants in terms of displacement field of the beam. Once the integrals are calculated, the following equation can be produced:

$$\begin{bmatrix} N \\ M \end{bmatrix} = \begin{bmatrix} A & B \\ B & D \end{bmatrix} \begin{bmatrix} \dfrac{\partial u}{\partial x} \\ \dfrac{\partial \varphi}{\partial x} \end{bmatrix},$$
$$Q = A^s \left(\varphi + \frac{\partial w}{\partial x} \right) \tag{2.100}$$

where

$$[A, B, D] = \int_A \left[1, z, z^2 \right] Q_{11} dA,$$
$$A^s = \kappa_s \int_A Q_{55} dA \tag{2.101}$$

Again, the terms Q_{ij} will be determined in Section 2.3 once the homogenization of different types of nanocomposites are discussed in detail. The governing equations can be derived now by substituting for the stress resultants from Eq. (2.100) in Eqs. (2.97)–(2.99). Thus, the governing equations can be reached as follows:

$$A \frac{\partial^2 u}{\partial x^2} + B \frac{\partial^2 \varphi}{\partial x^2} - I_0 \frac{\partial^2 u}{\partial t^2} - I_1 \frac{\partial^2 \varphi}{\partial t^2} = 0 \tag{2.102}$$

$$A^s \left(\frac{\partial^2 w}{\partial x^2} + \frac{\partial \varphi}{\partial x} \right) - N^T \frac{\partial^2 w}{\partial x^2} - I_0 \frac{\partial^2 w}{\partial t^2} = 0 \tag{2.103}$$

$$B\frac{\partial^2 u}{\partial x^2} + D\frac{\partial^2 \varphi}{\partial x^2} - A^s\left(\frac{\partial w}{\partial x} + \varphi\right) - I_1\frac{\partial^2 u}{\partial t^2} - I_2\frac{\partial^2 \varphi}{\partial t^2} = 0 \qquad (2.104)$$

In Eq. (2.103), the thermal force can be calculated using the following equation:

$$N^T = \int_A E_{11}(z)\alpha_{11}(z)(T - T_0)\,dA \qquad (2.105)$$

In previous equation, E_{11}, α_{11}, T, and T_0 are the Young's moduli in the axial direction, CTE in axial direction, local temperature, and reference temperature, respectively. The type of thermal loading depends on the profile used to present the temperature distribution. In other words, linear, sinusoidal, and nonlinear equations can be defined for the temperature field T as well as the simplest one (i.e., the uniform-type temperature raise). Now, solving the equations above reveals the mechanical response of the system. Solving Eqs. (2.102)–(2.104) determines the natural frequency of the Timoshenko beams. The critical buckling load of the beam can be obtained once the terms including differentiation with respect to time are omitted and a buckling load (i.e., N^b) is inserted in the governing equations, the same as thermal force. In this section, we present an efficient and powerful numerical approach for the purpose of solving the governing equations of Timoshenko nanocomposite beams. The method is one of the best transformation-based numerical methods to derive the mechanical response of the system.

In this approach, the differentiations with respect to the spatial independent variables will be calculated using their transformed equivalent expression. Due to this fact, this method is called the differential transformation method (DTM). This method enables us to make ordinary and partial differential equations algebraic ones. Before explaining the application of this method for static and dynamic problems, it is better to discuss the transformation of the kth order derivation of a desired function from its initial form to the mapped form. Assume a desired function $f(x)$, where its kth derivation with respect to the independent variable x is in the following form:

$$f^{(k)}(x) = \frac{d^k f(x)}{dx^k} \qquad (2.106)$$

The differential transformation of the kth derivative of function $f(x)$ and the differential inverse transformation of $F(k)$ can be defined as follows:

$$F(k) = \frac{1}{k!}\left[\frac{d^k}{dx^k}f(x)\right]_{x=0} \qquad (2.107)$$

$$f(x) = \sum_{k=0}^{\infty} x^k F(k) \qquad (2.108)$$

Substituting for the transformed function $F(k)$ from Eq. (2.107) in Eq. (2.108) results in the following expression for the original function:

$$f(x) = \sum_{k=0}^{\infty} \frac{x^k}{k!} \left[\frac{d^k}{dx^k} f(x) \right]_{x=0} \qquad (2.109)$$

Thus, it can be seen that the DTM originates from the concept of the Taylor's series expansion. It is noteworthy that in real applications, the infinite expansion is not needed to calculate the function $f(x)$. In fact, it is usual to select a finite number of terms, for example N, to reach the original function based on the transformed one. The reason for this selection is that from term number $N+1$ on, summation of all of the remained terms is small enough to be ignored. This can be summarized in a mathematical framework as follows:

$$f(x) = \sum_{k=0}^{N} \frac{x^k}{k!} \left[\frac{d^k}{dx^k} f(x) \right]_{x=0} \qquad (2.110)$$

The appropriate number of terms that result in a high precision response can be determined by checking the convergence of the answers. Before starting to solve the formulated problem, it is useful to mention the differential transformation of some of the algebraic equations which may be required to be known during the solution procedure. These transformations are shown in Table 2.1.

In Table 2.1, λ stands for a scalar and δ is the Dirac delta function. Now the dynamic responses of the problem can be obtained using the DTM. To start solving the problem, it is better to present the solution functions for the displacement field of the beam based on the concept of separation of variables. Based on this approach, the displacement field can be written in the following form:

$$\begin{aligned} u(x,t) &= u(x)e^{i\omega t}, \\ \varphi(x,t) &= \varphi(x)e^{i\omega t}, \\ w(x,t) &= w(x)e^{i\omega t} \end{aligned} \qquad (2.111)$$

TABLE 2.1

Transformation Rules in the One Dimensional (1-D) DTM

Original Function	Transformed Function
$f(x) = g(x) \pm h(x)$	$F(k) = G(k) \pm H(k)$
$f(x) = \lambda g(x)$	$F(k) = \lambda G(k)$
$f(x) = g(x)h(x)$	$F(k) = \sum_{l=0}^{k} G(k-l)H(l)$
$f(x) = \dfrac{d^n g(x)}{dx^n}$	$F(k) = \dfrac{(k+n)!}{k!} G(k+n)$
$f(x) = x^n$	$F(k) = \delta(k-n)$

---

where ω is the natural frequency. Inserting Eq. (2.111) in Eqs. (2.102)–(2.104), the following set of equations can be produced:

$$A\frac{\partial^2 u}{\partial x^2} + B\frac{\partial^2 \varphi}{\partial x^2} + I_0\omega^2 u + I_1\omega^2\varphi = 0 \tag{2.112}$$

$$A^s\left(\frac{\partial^2 w}{\partial x^2} + \frac{\partial\varphi}{\partial x}\right) - N^T\frac{\partial^2 w}{\partial x^2} + I_0\omega^2 w = 0 \tag{2.113}$$

$$B\frac{\partial^2 u}{\partial x^2} + D\frac{\partial^2 \varphi}{\partial x^2} - A^s\left(\frac{\partial w}{\partial x} + \varphi\right) + I_1\omega^2 u + I_2\omega^2\varphi = 0 \tag{2.114}$$

Regarding the rules explained in Table 2.1 and using the DTM, Eqs. (2.112)–(2.114) can be rewritten in the following form:

$$A(k+1)(k+2)U(k) + B(k+1)(k+2)\Phi(k) + I_0\omega^2 U(k) + I_1\omega^2\Phi(k) = 0 \tag{2.115}$$

$$A^s(k+1)(k+2)W(k) + A^s(k+1)\Phi(k) - N^T(k+1)(k+2)W(k) + I_0\omega^2 W(k) = 0 \tag{2.116}$$

$$B(k+1)(k+2)U(k) + D(k+1)(k+2)\Phi(k) - A^s(k+1)W(k)$$
$$- A^s\Phi(k) + I_1\omega^2 U(k) + I_2\omega^2\Phi(k) = 0 \tag{2.117}$$

where $U(k)$, $\Phi(k)$, and $W(k)$ are the transformed functions of $u(x)$, $\varphi(x)$, and $W(x)$, respectively. Now the algebraic set of the governing equations is obtained. To complete the solution procedure, the effects of BCs at the ends of the beam-type structure must be included, too.

Table 2.2 contains the transformations required to apply the effects of BCs at the edges of the beam. Based on this table, the mathematical representation of the simply supported BC at both ends of the beam can be written as follows:

$$U(0) = W(0) = W(2) = 0,$$

$$\sum_{k=0}^{\infty} U(k) = \sum_{k=0}^{\infty} W(k) = \sum_{k=0}^{\infty} k(k-1)W(k) = 0 \tag{2.118}$$

Using Eqs. (2.115)–(2.117) with the aid of the transformed BCs results in an eigenvalue problem. The mechanical response of the problem, either natural frequency or critical buckling load, can be obtained by solving the equation. Solving the equation is not too complex. In fact, the mechanical response can be derived by setting the determinant of the coefficient matrix to be zero. Recall that the critical buckling load of the problem can be achieved once the time-dependent terms are ignored. The convergence of the response can be checked by comparing the

TABLE 2.2

Original and Transformed Form of the BCs at the Ends of the Beam Based upon DTM

$x = 0$		$x = L$	
Original BC	**Transformed BC**	**Original BC**	**Transformed BC**
$f(0) = 0$	$F(0) = 0$	$f(L) = 0$	$\sum_{k=0}^{\infty} F(k) = 0$
$\dfrac{df(0)}{dx}$	$F(1) = 0$	$\dfrac{df(L)}{dx}$	$\sum_{k=0}^{\infty} kF(k) = 0$
$\dfrac{d^2 f(0)}{dx^2}$	$F(2) = 0$	$\dfrac{d^2 f(L)}{dx^2}$	$\sum_{k=0}^{\infty} k(k-1)F(k) = 0$
$\dfrac{d^3 f(0)}{dx^3}$	$F(3) = 0$	$\dfrac{d^3 f(L)}{dx^3}$	$\sum_{k=0}^{\infty} k(k-1)(k-2)F(k) = 0$

answers generated by changing the number of terms in the DTM. In other words, after deriving the first response, the other responses will be compared with their previous one and the process ends once the difference between two sequential responses is smaller than a very tiny value like ζ. This term can be assumed to be any desired scalar such as 1e-5 or smaller.

2.2.4 THIRD-ORDER SHEAR DEFORMATION BEAM THEORY

In this part of the book, the governing equations of a beam-type structure will be achieved using the well-known third-order shear deformation theory (TSDT) of Reddy. Similar to previous sections, the governing equations will be achieved for the dynamic problem and the motion equations of the static buckling problem can be enriched by eliminating the time-dependent terms. This theory presents a poly-nomial third-order approximation for the distribution of the shear stress and strain through the thickness of the beam. According to this theory, the displacement field of the beam can be presented in the following form:

$$u_x(x,z,t) = u(x,t) + z\varphi(x,t) - c_1 z^3 \left(\varphi(x,t) + \frac{\partial w(x,t)}{\partial x} \right),$$

$$u_z(x,z,t) = w(x,t) \tag{2.119}$$

where $c_1 = \frac{4}{3h^2}$. Also, u, φ, and w are axial displacement, rotation of the cross-section, and bending deflection, respectively. Now, the arrays of strain tensor can be derived using the equations above. Incorporating Eqs. (2.18) and (2.119), the nonzero strains of the beam can be produced in the following form:

$$\varepsilon_{xx} = \frac{1}{2}\left(\frac{\partial u_x}{\partial x} + \frac{\partial u_x}{\partial x}\right) = \frac{\partial u}{\partial x} + z\frac{\partial \varphi}{\partial x} - c_1 z^3\left(\frac{\partial \varphi}{\partial x} + \frac{\partial^2 w}{\partial x^2}\right),$$

$$\gamma_{xz} = 2\varepsilon_{xz} = \left(\frac{\partial u_x}{\partial z} + \frac{\partial u_z}{\partial x}\right) = \left(1 - c_2 z^2\right)\left(\varphi + \frac{\partial w}{\partial x}\right)$$

(2.120)

where $c_2 = 3c_1$. Again, ε_{xx} and γ_{xz} are normal strain in the axial direction and γ_{xz} is the shear strain in the xz plane. Now it is possible to start deriving the motion equations of the beam according to the Reddy's theory. As in previous sections, the concept of the dynamic form of the principle of virtual work will be implemented to reach the dynamic motion equations. These equations can then be reduced to the static case by eliminating the terms including the time variable itself or its derivatives. Now the variation of the strain energy of the beam can be formulated. The initial form of the variation of strain energy in Reddy's theory is completely as same as that introduced in Eq. (2.90), therefore, it will not be rewritten. Inserting the strain components from Eq. (2.120) in Eq. (2.90) reveals the following formula for the variation of the strain energy:

$$\delta U = \int_0^L \left[N\left(\delta\frac{\partial u}{\partial x}\right) + M\left(\delta\frac{\partial \varphi}{\partial x}\right) - c_1 P\left(\delta\left[\frac{\partial \varphi}{\partial x} + \frac{\partial^2 w}{\partial x^2}\right]\right)\right.$$
$$\left. + \left(Q - c_2 R\right)\left(\delta\left[\frac{\partial w}{\partial x} + \varphi\right]\right)\right] dx$$

(2.121)

in which the stress resultants utilized in the previous equation can be calculated using the following formula:

$$N = \int_A \sigma_{xx} dA,$$

$$M = \int_A z\sigma_{xx} dA,$$

$$P = \int_A z^3 \sigma_{xx} dA,$$

(2.122)

$$Q = \int_A \sigma_{xz} dA,$$

$$R = \int_A z^2 \sigma_{xz} dA$$

It can be observed that the TSDT is free from implementation of the shear correction factor, previously used in the FSDT. Next, the variation of the kinetic energy of the system must be formulated. The initial definition for the variation of the kinetic energy in the TSDT is similar with that presented in Eq. (2.93). To reach the expansion of this variation for a beam-type structure simulated based on the Reddy's beam model, Eq. (2.119) must be substituted in Eq. (2.93). Once this substitution is performed, the variation of the kinetic energy can be written as follows:

$$
\delta T = \int_0^L
\begin{bmatrix}
I_0\left(\dfrac{\partial u}{\partial t}\dfrac{\partial \delta u}{\partial t} + \dfrac{\partial w}{\partial t}\dfrac{\partial \delta w}{\partial t}\right) + I_1\left(\dfrac{\partial \varphi}{\partial t}\dfrac{\partial \delta u}{\partial t} + \dfrac{\partial u}{\partial t}\dfrac{\partial \delta \varphi}{\partial t}\right) + I_2\dfrac{\partial \varphi}{\partial t}\dfrac{\partial \delta \varphi}{\partial t} \\[4mm]
+c_1\left(
\begin{aligned}
&-I_3\dfrac{\partial u}{\partial t}\left[\dfrac{\partial^2 \delta w}{\partial x \partial t} + \dfrac{\partial \delta \varphi}{\partial t}\right] - I_3\dfrac{\partial \delta u}{\partial t}\left[\dfrac{\partial^2 w}{\partial x \partial t} + \dfrac{\partial \varphi}{\partial t}\right] \\[2mm]
&-I_4\dfrac{\partial \varphi}{\partial t}\left[\dfrac{\partial^2 \delta w}{\partial x \partial t} + \dfrac{\partial \delta \varphi}{\partial t}\right] - I_4\dfrac{\partial \delta \varphi}{\partial t}\left[\dfrac{\partial^2 w}{\partial x \partial t} + \dfrac{\partial \varphi}{\partial t}\right] \\[2mm]
&+c_1 I_6\left[\dfrac{\partial^2 w}{\partial x \partial t} + \dfrac{\partial \varphi}{\partial t}\right]\left[\dfrac{\partial^2 \delta w}{\partial x \partial t} + \dfrac{\partial \delta \varphi}{\partial t}\right]
\end{aligned}
\right)
\end{bmatrix}
dx \quad (2.123)
$$

where I_i, $i = (0,1,2,3,4,6)$ components employed in the equation above are the mass moments of inertia and can be computed using the following formula:

$$
I_i = \int_A z^i \rho(z) dA, \quad i = (0,1,2,3,4,6) \tag{2.124}
$$

The derivation procedure can be finished once the variation of the work done by external forces is presented in the mathematical framework. Now we will go about generating an expression for the variation of work done by external loading for a beam which is assumed to be embedded on a three-parameter visco-Pasternak substrate and placed in a thermally influenced environment. The elastic coefficients of the springs utilized in the medium are k_w and k_p for Winkler and Pasternak springs, respectively; also, the damping coefficient of the medium, introduced in the role of a linear damper, will be shown with c_d. In addition, the force applied on the beam due to the existence of thermal gradients will be shown with N^T. The variation of work done by medium and thermal loading can be expressed as follows:

$$
\delta V = \int_0^L\left(-k_w w + k_p \frac{\partial^2 w}{\partial x^2} - c_d \frac{\partial w}{\partial t} + N^T \frac{\partial^2 w}{\partial x^2}\right)\delta w\, dx \tag{2.125}
$$

Now the Euler-Lagrange equations of beams modeled by TSDT can be generated. First, the expressions introduced in Eqs. (2.121), (2.123), and (2.125) should be inserted in Eq. (2.21). Afterward, the coefficient multiplied by the variation of each of the displacement field's components must be assumed to be identical with zero in order to reach the nontrivial response of the generated equation. For more information, readers are asked to re-read Section 2.2.1. Doing the mathematical manipulation, the Euler-Lagrange equations can be derived in the following form:

$$\frac{\partial N}{\partial x} - I_0 \frac{\partial^2 u}{\partial t^2} - \hat{I}_1 \frac{\partial^2 \varphi}{\partial t^2} + c_1 I_3 \frac{\partial^3 w}{\partial x \partial t^2} = 0, \tag{2.126}$$

$$\frac{\partial \hat{M}}{\partial x} - \hat{Q} - \hat{I}_1 \frac{\partial^2 u}{\partial t^2} - \hat{I}_2 \frac{\partial^2 \varphi}{\partial t^2} + c_1 \hat{I}_4 \left(\frac{\partial^2 \varphi}{\partial t^2} + \frac{\partial^3 w}{\partial x \partial t^2} \right) = 0, \tag{2.127}$$

$$\frac{\partial \hat{Q}}{\partial x} + c_1 \frac{\partial^2 P}{\partial x^2} - k_w w + \left(k_p - N^T \right) \frac{\partial^2 w}{\partial x^2} - c_d \frac{\partial w}{\partial t} - I_0 \frac{\partial^2 w}{\partial t^2}$$
$$- c_1 I_3 \frac{\partial^3 u}{\partial x \partial t^2} - c_1 I_4 \frac{\partial^3 \varphi}{\partial x \partial t^2} + c_1^2 I_6 \left(\frac{\partial^3 \varphi}{\partial x \partial t^2} + \frac{\partial^4 w}{\partial x^2 \partial t^2} \right) \tag{2.128}$$

where

$$\hat{M} = M - c_1 P, \quad \hat{Q} = Q - c_2 R \tag{2.129}$$

and

$$\hat{I}_1 = I_1 - c_1 I_3, \quad \hat{I}_2 = I_2 - c_1 I_4, \quad \hat{I}_4 = I_4 - c_1 I_6 \tag{2.130}$$

The Euler-Lagrange equations are now completely derived and these equations must be presented in terms of the components of the displacement field to enrich the governing equations. Thus, the stress resultants must be expressed on the basis of the strain-displacement relationships defined in Eq. (2.120). To this purpose, integrals over the cross-section of the beam must be calculated to present a mathematical equivalent for the stress resultants in terms of displacement field. Once these mathematical operations are produced, the following relations can be generated:

$$\begin{bmatrix} N \\ M \\ P \end{bmatrix} = \begin{bmatrix} A & B - c_1 E & -c_1 E \\ B & D - c_1 F & -c_1 F \\ E & F - c_1 H & -c_1 H \end{bmatrix} \begin{bmatrix} \dfrac{\partial u}{\partial x} \\ \dfrac{\partial \varphi}{\partial x} \\ \dfrac{\partial^2 w}{\partial x^2} \end{bmatrix}, \tag{2.131}$$

$$\begin{bmatrix} Q \\ R \end{bmatrix} = \begin{bmatrix} A^s - c_2 D^s \\ D^s - c_2 F^s \end{bmatrix} \left(\varphi + \frac{\partial w}{\partial x} \right)$$

In the equation above, the cross-sectional rigidities can be calculated using the following relations:

$$[A, B, D, E, F, H] = \int_A [1, z, z^2, z^3, z^4, z^6] Q_{11} dA,$$

$$[A^s, D^s, F^s] = \int_A [1, z^2, z^4] Q_{55} dA$$

(2.132)

The governing equations of the problem can be generated once Eqs. (2.129)–(2.131) are inserted in Eqs. (2.126)–(2.128). After mathematical manipulations, the final governing equations of motion can be presented as follows:

$$A\frac{\partial^2 u}{\partial x^2} + (B - c_1 E)\frac{\partial^2 \varphi}{\partial x^2} - c_1 E\frac{\partial^3 w}{\partial x^3} - I_0\frac{\partial^2 u}{\partial t^2} - \hat{I}_1\frac{\partial^2 \varphi}{\partial t^2} + c_1 I_3\frac{\partial^3 w}{\partial x \partial t^2} = 0, \quad (2.133)$$

$$B\frac{\partial^2 u}{\partial x^2} + (D - c_1 F)\frac{\partial^2 \varphi}{\partial x^2} - c_1 F\frac{\partial^3 w}{\partial x^3}$$
$$- c_1 \left[E\frac{\partial^2 u}{\partial x^2} + (F - c_1 H)\frac{\partial^2 \varphi}{\partial x^2} - c_1 H\frac{\partial^3 w}{\partial x^3} \right]$$
$$- \left(A^s - c_2 D^s - c_2 [D^s - c_2 F^s] \right)\left(\varphi + \frac{\partial w}{\partial x} \right)$$
$$- \hat{I}_1\frac{\partial^2 u}{\partial t^2} - \hat{I}_2\frac{\partial^2 \varphi}{\partial t^2} + c_1 \hat{I}_4 \left(\frac{\partial^2 \varphi}{\partial t^2} + \frac{\partial^3 w}{\partial x \partial t^2} \right) = 0,$$

(2.134)

$$\left(A^s - c_2 D^s - c_2 [D^s - c_2 F^s] \right)\left(\frac{\partial \varphi}{\partial x} + \frac{\partial^2 w}{\partial x^2} \right)$$
$$+ c_1 \left[E\frac{\partial^3 u}{\partial x^3} + (F - c_1 H)\frac{\partial^3 \varphi}{\partial x^3} - c_1 H\frac{\partial^4 w}{\partial x^4} \right]$$
$$- k_w w + (k_p - N^T)\frac{\partial^2 w}{\partial x^2} - c_d\frac{\partial w}{\partial t} - I_0\frac{\partial^2 w}{\partial t^2} - c_1 I_3\frac{\partial^3 u}{\partial x \partial t^2}$$
$$- c_1 I_4\frac{\partial^3 \varphi}{\partial x \partial t^2} + c_1^2 I_6 \left(\frac{\partial^3 \varphi}{\partial x \partial t^2} + \frac{\partial^4 w}{\partial x^2 \partial t^2} \right)$$

(2.135)

Now the governing equations of the vibration problem of a third-order beam is in hand. Remember that the governing equations of the static buckling can be obtained by eliminating terms including differentiation with respect to time. Solving the equations results in finding the natural frequency of the beam-type elements. The well-known Navier's solution will be employed to solve the governing equations once both ends of the beam are simply supported. Based on this solution, the displacement field of the beam can be assumed to have the following solution function for a dynamic analysis:

$$u = \sum_{n=1}^{\infty} U_n \cos\left(\frac{n\pi x}{L}\right) e^{i\omega_n t},$$

$$w = \sum_{n=1}^{\infty} W_n \sin\left(\frac{n\pi x}{L}\right) e^{i\omega_n t}, \tag{2.136}$$

$$\varphi = \sum_{n=1}^{\infty} \Phi_n \cos\left(\frac{n\pi x}{L}\right) e^{i\omega_n t}$$

where U_n, W_n, and Φ_n are the unknown amplitudes and the term ω_n indicates the natural frequency of the system. Once Eq. (2.136) is substituted into Eqs. (2.133)–(2.135), an eigenvalue equation similar with that introduced in Eq. (2.83) can be attained. However, the equation in this section will be a 3×3 eigenvalue equation. Going through the instructions presented in Section 2.2.1, the final response of the problem (i.e., the natural frequency of the beam) can be developed. Again, it must be stated that the critical buckling load of the beam in the static state can be extracted by eliminating time-dependent terms and also considering a buckling load applied on the beam just as same as thermal force. Clearly, the solution functions for the static buckling problem do not include the exponential time-dependent term of $e^{i\omega_n t}$.

2.2.5   REFINED HIGHER-ORDER BEAM THEORY

In this section, the expansion of the well-known Hamilton's principle will be presented for beam-type elements modeled via a refined shear deformation beam hypothesis. Such theories can estimate the mechanical response of beams free from any shear correction coefficient the same as simple higher-order beam models. However, there is a small difference between refined and simple higher-order beam theories. In other words, in refined theories, no parameter will be employed for the rotation of the beam's cross-section and the deflection of the beam will be divided in two terms indicating bending and shear deflections, respectively. Thus it is easier to formulate the problem according to refined models rather than conventional ones. Again, it is worth mentioning that the derivation of the motion equations will be performed for the case of analyzing a dynamic problem and the motion equations of the static buckling problem can be easily achieved by dismissing the terms including time or derivatives with respect to it. Now, the displacement field of a refined shear deformable beam can be presented as follows:

$$u_x(x,z,t) = u(x,t) - z\frac{\partial w_b(x,t)}{\partial x} - f(z)\frac{\partial w_s(x,t)}{\partial x}, \tag{2.137}$$

$$u_z(x,z,t) = w_b(x,t) + w_s(x,t)$$

where u is the axial displacement; also, w_b and w_s are bending and shear deflections, respectively. In the equation above, $f(z)$ is a shape function presented to govern the distribution of shear stress and strain across the beam's thickness. This function can be trigonometric, polynomial, or exponential. In this study, we will use the well-known third-order function to be the shape function. Thus the shape function can be expressed by:

$$f(z) = z - \frac{4z^3}{3h^2} \tag{2.138}$$

Now, the strain-displacement relations can be produced using the displacement field introduced in Eq. (2.137) incorporated with the definition of the Eulerian strain tensor (see Eq. (2.18)). Once these equations are combined, the nonzero components of the strain tensor can be written in the following form:

$$\varepsilon_{xx} = \frac{1}{2}\left(\frac{\partial u_x}{\partial x} + \frac{\partial u_x}{\partial x}\right) = \frac{\partial u}{\partial x} + z\frac{\partial^2 w_b}{\partial x^2} - f(z)\frac{\partial^2 w_s}{\partial x^2},$$

$$\gamma_{xz} = 2\varepsilon_{xz} = \left(\frac{\partial u_x}{\partial z} + \frac{\partial u_z}{\partial x}\right) = g(z)\frac{\partial w_s}{\partial x} \tag{2.139}$$

in which

$$g(z) = 1 - \frac{df(z)}{dz} \tag{2.140}$$

The energy-based approach can be used to produce the governing equations for refined shear deformable beams. Herein, the concept of the dynamic form of the principle of virtual work will be implemented to reach the motion equations of beams. Actually, the variation of the strain energy of refined shear deformable beams is entirely as same as the expression presented in Eq. (2.90). Once the strain-displacement relations of the employed beam theory (i.e., Eq. (2.139)) is inserted in Eq. (2.90), the variation of the strain energy of the beam can be presented in the following form:

$$\delta U = \int_0^L \left[N\left(\delta\frac{\partial u}{\partial x}\right) - M^b\left(\delta\frac{\partial^2 w_b}{\partial x^2}\right) - M^s\left(\delta\frac{\partial^2 w_s}{\partial x^2}\right) + Q\left(\delta\frac{\partial w_s}{\partial x}\right) \right] dx \tag{2.141}$$

where the stress resultants used in the equation above can be defined in the following form:

$$\left[N, M^b, M^s\right] = \int_A \left[1, z, f(z)\right]\sigma_{xx} dA,$$

$$Q = \int_A g(z)\sigma_{xz} dA \tag{2.142}$$

Now the variation of the kinetic energy of the beam must be derived. The initial definition of the variation of the kinetic energy for a refined higher-order beam is as same as that reported previously in Eq. (2.93). The variation of the kinetic energy can be obtained for a refined beam by substituting for the displacement field of the beam from Eq. (2.137) into Eq. (2.93). Once this manipulation is completed, the variation of the kinetic energy of the beam will be in the following form:

$$
\delta T = \int_0^L \left[\begin{array}{l} I_0 \left(\dfrac{\partial u}{\partial t} \dfrac{\partial \delta u}{\partial t} + \dfrac{\partial (w_b + w_s)}{\partial t} \dfrac{\partial \delta (w_b + w_s)}{\partial t} \right) - I_1 \left(\dfrac{\partial u}{\partial t} \dfrac{\partial^2 \delta w_b}{\partial x \partial t} + \dfrac{\partial \delta u}{\partial t} \dfrac{\partial^2 w_b}{\partial x \partial t} \right) \\[3mm] -J_1 \left(\dfrac{\partial u}{\partial t} \dfrac{\partial^2 \delta w_s}{\partial x \partial t} + \dfrac{\partial \delta u}{\partial t} \dfrac{\partial^2 w_s}{\partial x \partial t} \right) + I_2 \dfrac{\partial^2 w_b}{\partial x \partial t} \dfrac{\partial^2 \delta w_b}{\partial x \partial t} + K_2 \dfrac{\partial^2 w_s}{\partial x \partial t} \dfrac{\partial^2 \delta w_s}{\partial x \partial t} \\[3mm] + J_2 \left(\dfrac{\partial^2 w_b}{\partial x \partial t} \dfrac{\partial^2 \delta w_s}{\partial x \partial t} + \dfrac{\partial^2 w_s}{\partial x \partial t} \dfrac{\partial^2 \delta w_b}{\partial x \partial t} \right) \end{array} \right] dx \quad (2.143)
$$

where the mass moments of inertial employed in the equation above can be calculated using the following formula:

$$
[I_0, I_1, J_1, I_2, J_2, K_2] = \int_A \left[1, z, f(z), z^2, zf(z), f^2(z) \right] \rho(z) dA \quad (2.144)
$$

The Euler-Lagrange equations of the beam can be derived by presenting an expression for the work done by external forces. The beam is assumed to be subjected to both thermally and magnetically influenced environments. Therefore, the effects of external loading applied on the structure must be considered regarding the magnetic field intensity and thermal gradients. Two ways to apply magnetic force on the beam will be discussed. First, a unidirectional simple magnetic field with uniform distribution will be used. Alternately, a sinusoidal magnetic field can be employed to account for the changes which can happen in the intensity of the magnetic field in each magnetically affected environment.

Assume a magnetic field with the following intensity vector:

$$
\mathbf{H} = (H_x, 0, 0) \quad (2.145)
$$

Using the equation above for the applied unidirectional magnetic field and also expanding the Maxwell's magnetic induction relations, the external force applied on the beam due to the existence of external magnetic field can be formulated in the following form:

$$
f_{Lz} = \eta \left(\underbrace{\overbrace{\nabla \times (\nabla \times (\mathbf{u} \times \mathbf{H}))}^{\mathbf{J}}}_{\mathbf{h}} \right) \times \mathbf{H} \quad (2.146)
$$

The force applied on the structure is called Lorentz force and is shown with the sign f_{Lz}. In the equation above, $\mathbf{u}$ is the displacement field vector and $\mathbf{H}$ is the vector of the magnetic field's intensity. Moreover, ∇ is gradient operator; also, $\mathbf{J}$, $\mathbf{h}$, and η denote the current density vector, the vector of applied magnetic field, and magnetic permeability, respectively. Now, the final body force applied on the structure can be determined once the displacement field of refined beam model (i.e., Eq. (2.137)) is substituted in Eq. (2.146). Therefore, the body Lorentz force applied on the beam-type element can be expressed in the following form:

$$f_{Lz} = \eta H_x^2 \frac{\partial^2 (w_b + w_s)}{\partial x^2} \tag{2.147}$$

The variation of the work done by external loading can be formulated for the case of placing the beam in an environment with a unidirectional magnetic field. The variation of the work done by external loading in this condition can be presented in the following form:

$$\delta V = \int_0^L \left(N^T \frac{\partial^2 (w_b + w_s)}{\partial x^2} - F_{Lz} \right) \delta(w_b + w_s)\, dx \tag{2.148}$$

where

$$F_{Lz} = \int_A f_{Lz}\, dA \tag{2.149}$$

Also, the thermal force applied on the beam can be considered to be as same as that defined in Eq. (2.105). Prior to presenting the Euler-Lagrange equations of the beam, it is better to discuss about the impact of the non-uniform sinusoidal magnetic field on the beam-type elements. In this case, the vector of the magnetic field's intensity is again as same as Eq. (2.145); however, the magnitude of the intensity is not uniform all over the beam's length. In other words, the magnetic field's intensity possesses its peak value at the middle of the beam and is identical with zero at the ends. Therefore, this parameter can be defined in the following form:

$$H_x = \bar{H}_x \sin\left(\frac{\pi x}{L} \right) \tag{2.150}$$

where $\bar{H}_x$ denotes the amplitude of the magnetic field intensity. Once Eqs. (2.137), (2.146), and (2.150) are combined, the following expression will be produced for the applied magnetic force on the beam:

$$f_{Lz} = \eta \left(H_x^2 \frac{\partial^2 (w_b + w_s)}{\partial x^2} + 2H_x \frac{\partial H_x}{\partial x} \frac{\partial (w_b + w_s)}{\partial x} + H_x \frac{\partial^2 H_x}{\partial x^2} (w_b + w_s) \right) \tag{2.151}$$

Integrating from the previous equation over the cross-sectional area of the beam results in the following expression:

$$F_{Lz} = \int_A f_{Lz} dA = \psi_1 \frac{\partial^2 (w_b + w_s)}{\partial x^2} + \psi_2 \frac{\partial (w_b + w_s)}{\partial x} - \psi_3 (w_b + w_s) \quad (2.152)$$

where

$$\psi_1 = \eta A \bar{H}_x^2 \sin^2 \left(\frac{\pi x}{L} \right),$$

$$\psi_2 = 2\eta A \left(\frac{\pi}{L} \right) \bar{H}_x^2 \sin \left(\frac{\pi x}{L} \right) \cos \left(\frac{\pi x}{L} \right), \quad (2.153)$$

$$\psi_3 = \eta A \left(\frac{\pi}{L} \right)^2 \bar{H}_x^2 \sin^2 \left(\frac{\pi x}{L} \right)$$

So, the work done by external forces for a beam subjected to thermal environment and placed in a non-uniform longitudinal magnetic field can be expressed in the following form:

$$\delta V = \int_0^L \left(N^T \frac{\partial^2 w}{\partial x^2} - \psi_1 \frac{\partial^2 (w_b + w_s)}{\partial x^2} - \psi_2 \frac{\partial (w_b + w_s)}{\partial x} \right.$$

$$\left. + \psi_3 (w_b + w_s) \right) \delta(w_b + w_s) dx \quad (2.154)$$

The Euler-Lagrange equations of the beam can be derived substituting for the variations of strain energy, kinetic energy, and work done by external loading in the definition of Hamilton's principle. Due to the fact that two types of magnetic field are implemented, two sets of Euler-Lagrange equations will be presented in what follows. First, the Euler-Lagrange equations of beams subjected to uniform magnetic field must be obtained by inserting Eqs. (2.141), (2.143), and (2.148) in Eq. (2.21). After completing the substitution and choosing the nontrivial response of the equation, the Euler-Lagrange equations of the beam can be shown in the following form:

$$\frac{\partial N}{\partial x} - I_0 \frac{\partial^2 u}{\partial t^2} + I_1 \frac{\partial^3 w_b}{\partial x \partial t^2} + J_1 \frac{\partial^3 w_s}{\partial x \partial t^2} = 0, \quad (2.155)$$

$$\frac{\partial^2 M^b}{\partial x^2} + F_{Lz} - N^T \frac{\partial^2 (w_b + w_s)}{\partial x^2} - I_0 \frac{\partial^2 (w_b + w_s)}{\partial t^2}$$

$$-I_1 \frac{\partial^3 u}{\partial x \partial t^2} + I_2 \frac{\partial^4 w_b}{\partial x^2 \partial t^2} + J_2 \frac{\partial^4 w_s}{\partial x^2 \partial t^2} = 0, \quad (2.156)$$

$$\frac{\partial^2 M^s}{\partial x^2} + \frac{\partial Q}{\partial x} + F_{Lz} - N^T \frac{\partial^2 (w_b + w_s)}{\partial x^2} - I_0 \frac{\partial^2 (w_b + w_s)}{\partial t^2}$$

$$-J_1 \frac{\partial^3 u}{\partial x \partial t^2} + J_2 \frac{\partial^4 w_b}{\partial x^2 \partial t^2} + K_2 \frac{\partial^4 w_s}{\partial x^2 \partial t^2} = 0 \tag{2.157}$$

In the equations above, the term F_{Lz} can be employed from Eq. (2.149). Now, the Euler-Lagrange equations of beams subjected to thermal loading and non-uniform magnetic field can be obtained by substituting Eqs. (2.141), (2.143), and (2.154) in Eq. (2.21). Similar to previous problems, herein, the Euler-Lagrange equations can be produced by implementing the nontrivial response of the problem. After mathematical manipulations, the Euler-Lagrange equations will be as follows:

$$\frac{\partial N}{\partial x} - I_0 \frac{\partial^2 u}{\partial t^2} + I_1 \frac{\partial^3 w_b}{\partial x \partial t^2} + J_1 \frac{\partial^3 w_s}{\partial x \partial t^2} = 0, \tag{2.158}$$

$$\frac{\partial^2 M^b}{\partial x^2} + \psi_1 \frac{\partial^2 (w_b + w_s)}{\partial x^2} + \psi_2 \frac{\partial (w_b + w_s)}{\partial x}$$

$$-\psi_3 (w_b + w_s) - N^T \frac{\partial^2 (w_b + w_s)}{\partial x^2} - I_0 \frac{\partial^2 (w_b + w_s)}{\partial t^2} \tag{2.159}$$

$$-I_1 \frac{\partial^3 u}{\partial x \partial t^2} + I_2 \frac{\partial^4 w_b}{\partial x^2 \partial t^2} + J_2 \frac{\partial^4 w_s}{\partial x^2 \partial t^2} = 0,$$

$$\frac{\partial^2 M^s}{\partial x^2} + \frac{\partial Q}{\partial x} + \psi_1 \frac{\partial^2 (w_b + w_s)}{\partial x^2} + \psi_2 \frac{\partial (w_b + w_s)}{\partial x}$$

$$-\psi_3 (w_b + w_s) - N^T \frac{\partial^2 (w_b + w_s)}{\partial x^2} - I_0 \frac{\partial^2 (w_b + w_s)}{\partial t^2} \tag{2.160}$$

$$-J_1 \frac{\partial^3 u}{\partial x \partial t^2} + J_2 \frac{\partial^4 w_b}{\partial x^2 \partial t^2} + K_2 \frac{\partial^4 w_s}{\partial x^2 \partial t^2} = 0$$

Now the final governing equations must be derived using the stress-strain relationship. To reach the governing equations of the problem, some integrals over the cross-sectional area of the beam must be calculated to present the stress resultants in terms of the displacement field of the beam. Procuring the mathematical operations, the following formulas can be developed:

$$\begin{bmatrix} N \\ M^b \\ M^s \end{bmatrix} = \begin{bmatrix} A & B & B^s \\ B & D & D^s \\ B^s & D^s & H^s \end{bmatrix} \begin{bmatrix} \dfrac{\partial u}{\partial x} \\ -\dfrac{\partial^2 w_b}{\partial x^2} \\ -\dfrac{\partial^2 w_s}{\partial x^2} \end{bmatrix}, \tag{2.161}$$

$$Q = A^s \frac{\partial w_s}{\partial x}$$

In the equation above, the cross-sectional rigidities can be reached using the following formulas:

$$\left[A, B, D, B^s, D^s, H^s\right] = \int_A \left[1, z, z^2, f(z), zf(z), f^2(z)\right] Q_{11} dA,$$

$$A^s = \int_A g^2(z) Q_{55} dA \tag{2.162}$$

Now, the governing motion equations can be easily achieved by substituting for the stress resultants from Eq. (2.161) in the Euler-Lagrange equations. Once Eq. (2.161) is inserted in Eqs. (2.155)–(2.157), the governing equations can be written in the following form:

$$A \frac{\partial^2 u}{\partial x^2} - B \frac{\partial^3 w_b}{\partial x^3} - B^s \frac{\partial^3 w_s}{\partial x^3} - I_0 \frac{\partial^2 u}{\partial t^2} + I_1 \frac{\partial^3 w_b}{\partial x \partial t^2} + J_1 \frac{\partial^3 w_s}{\partial x \partial t^2} = 0, \tag{2.163}$$

$$B \frac{\partial^3 u}{\partial x^3} - D \frac{\partial^4 w_b}{\partial x^4} - D^s \frac{\partial^4 w_s}{\partial x^4} + \left(F_{Lz} - N^T\right) \frac{\partial^2 (w_b + w_s)}{\partial x^2} - I_0 \frac{\partial^2 (w_b + w_s)}{\partial t^2}$$

$$-I_1 \frac{\partial^3 u}{\partial x \partial t^2} + I_2 \frac{\partial^4 w_b}{\partial x^2 \partial t^2} + J_2 \frac{\partial^4 w_s}{\partial x^2 \partial t^2} = 0, \tag{2.164}$$

$$B^s \frac{\partial^3 u}{\partial x^3} - D^s \frac{\partial^4 w_b}{\partial x^4} - H^s \frac{\partial^4 w_s}{\partial x^4} + A^s \frac{\partial^2 w_s}{\partial x^2}$$

$$+\left(F_{Lz} - N^T\right) \frac{\partial^2 (w_b + w_s)}{\partial x^2} - I_0 \frac{\partial^2 (w_b + w_s)}{\partial t^2} \tag{2.165}$$

$$-J_1 \frac{\partial^3 u}{\partial x \partial t^2} + J_2 \frac{\partial^4 w_b}{\partial x^2 \partial t^2} + K_2 \frac{\partial^4 w_s}{\partial x^2 \partial t^2} = 0,$$

The governing equations of beams subjected to non-uniform magnetic field and thermal gradient can be achieved by substituting Eq. (2.161) in Eqs. (2.158)–(2.160). After the substitution, the governing equations of the beam under defined loading condition are as follows:

$$A \frac{\partial^2 u}{\partial x^2} - B \frac{\partial^3 w_b}{\partial x^3} - B^s \frac{\partial^3 w_s}{\partial x^3} - I_0 \frac{\partial^2 u}{\partial t^2} + I_1 \frac{\partial^3 w_b}{\partial x \partial t^2} + J_1 \frac{\partial^3 w_s}{\partial x \partial t^2} = 0, \tag{2.166}$$

$$B\frac{\partial^3 u}{\partial x^3} - D\frac{\partial^4 w_b}{\partial x^4} - D^s\frac{\partial^4 w_s}{\partial x^4} + \psi_1\frac{\partial^2(w_b + w_s)}{\partial x^2} + \psi_2\frac{\partial(w_b + w_s)}{\partial x} - \psi_3(w_b + w_s)$$

$$-N^T\frac{\partial^2(w_b + w_s)}{\partial x^2} - I_0\frac{\partial^2(w_b + w_s)}{\partial t^2} - I_1\frac{\partial^3 u}{\partial x \partial t^2} + I_2\frac{\partial^4 w_b}{\partial x^2 \partial t^2} + J_2\frac{\partial^4 w_s}{\partial x^2 \partial t^2} = 0, \tag{2.167}$$

$$B^s\frac{\partial^3 u}{\partial x^3} - D^s\frac{\partial^4 w_b}{\partial x^4} - H^s\frac{\partial^4 w_s}{\partial x^4} + A^s\frac{\partial^2 w_s}{\partial x^2} + \psi_1\frac{\partial^2(w_b + w_s)}{\partial x^2}$$

$$+\psi_2\frac{\partial(w_b + w_s)}{\partial x} - \psi_3(w_b + w_s) - N^T\frac{\partial^2(w_b + w_s)}{\partial x^2} \tag{2.168}$$

$$-I_0\frac{\partial^2(w_b + w_s)}{\partial t^2} - J_1\frac{\partial^3 u}{\partial x \partial t^2} + J_2\frac{\partial^4 w_b}{\partial x^2 \partial t^2} + K_2\frac{\partial^4 w_s}{\partial x^2 \partial t^2} = 0,$$

The previous sets of governing equations must be solved to reach the mechanical response of the system. It is worth mentioning that the governing equations of the static buckling problem can be reached by eliminating the terms which are associated with the time itself or its derivatives. Moreover, a static buckling load can be applied on the beam-type structure the same as the thermal force applied on the structure. Due to the fact that in one of the cases, the longitudinal magnetic field depends on the spatial coordinate system, the well-known Galerkin's method, as one of the most powerful analytical methods, will be implemented to derive the natural frequency and critical buckling load of the beam. Based on this method, the solution functions for the dynamic case can be presented in the following form:

$$u = \sum_{n=1}^{\infty} U_n \frac{\partial X_n(x)}{\partial x} e^{i\omega_n t},$$

$$w_b = \sum_{n=1}^{\infty} W_{bn} X_n(x) e^{i\omega_n t}, \tag{2.169}$$

$$w_s = \sum_{n=1}^{\infty} W_{sn} X_n(x) e^{i\omega_n t}$$

In the equation above, the term $X_n(x)$ is responsible for satisfying the BCs at the ends of the beam. Also, ω_n reveals the natural frequency of the beam in dynamic problems and it must be neglected once the static buckling problem is under observation. Substituting Eq. (2.169) in the governing equations results in reaching an eigenvalue equation similar to that introduced in Eq. (2.83). However, in this case there is no damping factor in the governing equations and a simple free vibration problem will be reached (i.e., the damping matrix equals with zero). The dimension of the final eigenvalue equation is 3×3 in this chapter. Finally, solving the achieved eigenvalue equation reveals the natural frequency of the system. It is worth mentioning that the static buckling problem of the beam can be solved ignoring time-dependent terms in both solution function and governing equations of the problem. The $X_n(x)$ functions for various types of BCs can be presented in the following form:

$$
\text{S-S: } X_n(x) = \sin\left(\frac{n\pi x}{L}\right),
$$

$$
\text{C-S: } X_n(x) = \sin\left(\frac{n\pi x}{L}\right)\left[\cos\left(\frac{n\pi x}{L}\right) - 1\right], \tag{2.170}
$$

$$
\text{C-C: } X_n(x) = \sin^2\left(\frac{n\pi x}{L}\right)
$$

2.2.6 THIRD-ORDER SHEAR DEFORMATION PLATE THEORY

In this section, the kinematic relations of the TSDT will be expanded for plate-type elements. The TSDT is able to capture the effects of shear deformation via a polynomial third-order function free from any external shear correction coefficient. In what follows, we will present a time-dependent dynamic formulation for the governing equations of plates. The governing equations produced can be transferred to the static state to be usable for the static buckling analysis of the plates. According to the TSDT of plates, the displacement field of a plate-type structure can be presented as follows:

$$
u_x(x,y,z,t) = u(x,y,t) + z\varphi_x(x,y,t) - c_1 z^3 \left(\varphi_x(x,y,t) + \frac{\partial w(x,y,t)}{\partial x} \right),
$$

$$
u_y(x,y,z,t) = v(x,y,t) + z\varphi_y(x,y,t) - c_1 z^3 \left(\varphi_y(x,y,t) + \frac{\partial w(x,y,t)}{\partial y} \right), \tag{2.171}
$$

$$
u_z(x,y,z,t) = w(x,y,t)
$$

where $c_1 = \frac{4}{3h^2}$. Also, u, v, and w are axial displacement, transverse displacement, and bending deflection, respectively. Moreover, the rotations of the normal to the mid-plane around y and x directions are shown with φ_x and φ_y, respectively. Now the strain-displacement relationships can be obtained for TSDT plates. To this purpose, the displacement field of this type of plates (see Eq. (2.171)) must be incorporated into the definition of the Eulerian strain tensor, presented in Eq. (2.18). Once these equations are combined, the nonzero strains of the plate can be derived in the following form:

$$\varepsilon_{xx} = \frac{1}{2}\left(\frac{\partial u_x}{\partial x} + \frac{\partial u_x}{\partial x}\right) = \frac{\partial u}{\partial x} + z\frac{\partial \varphi_x}{\partial x} - c_1 z^3\left(\frac{\partial \varphi_x}{\partial x} + \frac{\partial^2 w}{\partial x^2}\right),$$

$$\varepsilon_{yy} = \frac{1}{2}\left(\frac{\partial u_y}{\partial y} + \frac{\partial u_y}{\partial y}\right) = \frac{\partial v}{\partial y} + z\frac{\partial \varphi_y}{\partial y} - c_1 z^3\left(\frac{\partial \varphi_y}{\partial y} + \frac{\partial^2 w}{\partial y^2}\right),$$

$$\gamma_{xy} = 2\varepsilon_{xy} = \left(\frac{\partial u_x}{\partial y} + \frac{\partial u_y}{\partial x}\right) = \frac{\partial u}{\partial y} + \frac{\partial v}{\partial x} + z\left(\frac{\partial \varphi_x}{\partial y} + \frac{\partial \varphi_y}{\partial x}\right)$$

$$\quad\quad -c_1 z^3\left(\frac{\partial \varphi_x}{\partial y} + \frac{\partial \varphi_y}{\partial x} + 2\frac{\partial^2 w}{\partial x \partial y}\right),$$

$$\gamma_{xz} = 2\varepsilon_{xz} = \left(\frac{\partial u_x}{\partial z} + \frac{\partial u_z}{\partial x}\right) = \left(1 - c_2 z^2\right)\left(\varphi_x + \frac{\partial w}{\partial x}\right),$$

$$\gamma_{yz} = 2\varepsilon_{yz} = \left(\frac{\partial u_y}{\partial z} + \frac{\partial u_z}{\partial y}\right) = \left(1 - c_2 z^2\right)\left(\varphi_y + \frac{\partial w}{\partial y}\right)$$

(2.172)

Now one can start to derive the Euler-Lagrange equations of the plate on the basis of the strain-displacement relations achieved in the equation above. First, the variation of the strain energy of the plate must be derived. For plate-type structures, the definition of the strain energy can be assumed to be as follows:

$$\delta U = \int_{\forall} \sigma_{ij}\delta\varepsilon_{ij} d\forall = \int_{\forall}\left(\sigma_{xx}\delta\varepsilon_{xx} + \sigma_{yy}\delta\varepsilon_{yy} + \sigma_{xy}\delta\gamma_{xy} + \sigma_{xz}\delta\gamma_{xz} + \sigma_{yz}\delta\gamma_{yz}\right)d\forall \quad (2.173)$$

where $\forall$ stands for the total volume of the media. The equation above can be rewritten in the following form once Eq. (2.172) is inserted into it:

$$\delta U = \int_A \begin{bmatrix} N_{xx}\left(\delta\frac{\partial u}{\partial x}\right) + M_{xx}\left(\delta\frac{\partial\varphi_x}{\partial x}\right) - c_1 P_{xx}\left(\delta\left[\frac{\partial\varphi_x}{\partial x} + \frac{\partial^2 w}{\partial x^2}\right]\right) + N_{yy}\left(\delta\frac{\partial v}{\partial y}\right) \\[2mm] + M_{xx}\left(\delta\frac{\partial\varphi_y}{\partial y}\right) - c_1 P_{yy}\left(\delta\left[\frac{\partial\varphi_y}{\partial y} + \frac{\partial^2 w}{\partial y^2}\right]\right) + N_{xy}\left(\delta\left[\frac{\partial u}{\partial y} + \frac{\partial v}{\partial x}\right]\right) \\[2mm] + M_{xy}\left(\delta\left[\frac{\partial\varphi_x}{\partial y} + \frac{\partial\varphi_y}{\partial x}\right]\right) - c_1 P_{xy}\left(\delta\left[\frac{\partial\varphi_x}{\partial y} + \frac{\partial\varphi_y}{\partial x} + 2\frac{\partial^2 w}{\partial x\partial y}\right]\right) \\[2mm] + \left(Q_{xz} - c_2 S_{xz}\right)\left(\delta\left[\varphi_x + \frac{\partial w}{\partial x}\right]\right) + \left(Q_{yz} - c_2 S_{yz}\right)\left(\delta\left[\varphi_y + \frac{\partial w}{\partial y}\right]\right) \end{bmatrix} dA \quad (2.174)$$

in which the previous integration is fulfilled over the cross-sectional area of the plate constructed from the intersection of x and y directions (i.e., $\int_A (.)\,dA = \int_0^b \int_0^a (.)\,dx\,dy$). In the equation above, the stress resultants can be defined as follows:

$$\left[N_{xx}, N_{yy}, N_{xy} \right] = \int_{-\frac{h}{2}}^{\frac{h}{2}} \left[\sigma_{xx}, \sigma_{yy}, \sigma_{xy} \right] dz,$$

$$\left[M_{xx}, M_{yy}, M_{xy} \right] = \int_{-\frac{h}{2}}^{\frac{h}{2}} z \left[\sigma_{xx}, \sigma_{yy}, \sigma_{xy} \right] dz,$$

$$\left[P_{xx}, P_{yy}, P_{xy} \right] = \int_{-\frac{h}{2}}^{\frac{h}{2}} z^3 \left[\sigma_{xx}, \sigma_{yy}, \sigma_{xy} \right] dz,$$

$$\left[Q_{xz}, Q_{yz} \right] = \int_{-\frac{h}{2}}^{\frac{h}{2}} \left[\sigma_{xz}, \sigma_{yz} \right] dz,$$

$$\left[S_{xz}, S_{yz} \right] = \int_{-\frac{h}{2}}^{\frac{h}{2}} z^2 \left[\sigma_{xz}, \sigma_{yz} \right] dz$$

(2.175)

In addition, the variation of the kinetic energy of the plate should be formulized now. To this purpose, the definition of the variation of the kinetic energy must be rewritten here. Indeed, the formerly implemented holds only for shear deformable beams and it is not applicable for the dynamic analysis of a higher-order plate (see Eq. (2.93)). The variation of the kinetic energy can be presented as follows for a TSD plate:

$$\delta T = \int_\forall \rho(z) \left[\frac{\partial u_x}{\partial t} \frac{\partial \delta u_x}{\partial t} + \frac{\partial u_y}{\partial t} \frac{\partial \delta u_y}{\partial t} + \frac{\partial u_z}{\partial t} \frac{\partial \delta u_z}{\partial t} \right] d\forall \qquad (2.176)$$

The term must be expanded for a plate simulated via the TSDT. To this purpose, the displacement field of the plate must be substituted in Eq. (2.176). Once this substitution is carried out and mathematical manipulations are accomplished, the variation of the kinetic energy of the plate can be depicted as follows:

$$
\delta T = \int_A \left[
\begin{array}{l}
\left(I_0 \dfrac{\partial^2 u}{\partial t^2} - c_1 I_3 \dfrac{\partial^3 w}{\partial x \partial t^2} + [I_1 - c_1 I_3] \dfrac{\partial^2 \varphi_x}{\partial t^2} \right) \delta u \\[3mm]
+ \left(I_0 \dfrac{\partial^2 v}{\partial t^2} - c_1 I_3 \dfrac{\partial^3 w}{\partial y \partial t^2} + [I_1 - c_1 I_3] \dfrac{\partial^2 \varphi_y}{\partial t^2} \right) \delta v \\[3mm]
+ \left(
\begin{array}{l}
-c_1 I_3 \dfrac{\partial^3 u}{\partial x \partial t^2} - c_1 I_3 \dfrac{\partial^3 v}{\partial y \partial t^2} + I_0 \dfrac{\partial^2 w}{\partial t^2} + 2c_1{}^2 I_6 \dfrac{\partial^4 w}{\partial x^2 \partial t^2} \\[3mm]
+ c_1 [c_1 I_6 - I_4] \dfrac{\partial^3 \varphi_x}{\partial x \partial t^2} + c_1 [c_1 I_6 - I_4] \dfrac{\partial^3 \varphi_y}{\partial y \partial t^2}
\end{array}
\right) \delta w \\[3mm]
+ \left([I_1 - c_1 I_3] \dfrac{\partial^2 u}{\partial t^2} + c_1 [c_1 I_6 - I_4] \dfrac{\partial^3 w}{\partial x \partial t^2} + [I_2 - 2c_1 I_4 + c_1{}^2 I_6] \dfrac{\partial^2 \varphi_x}{\partial t^2} \right) \delta \varphi_x \\[3mm]
+ \left([I_1 - c_1 I_3] \dfrac{\partial^2 v}{\partial t^2} + c_1 [c_1 I_6 - I_4] \dfrac{\partial^3 w}{\partial y \partial t^2} + [I_2 - 2c_1 I_4 + c_1{}^2 I_6] \dfrac{\partial^2 \varphi_y}{\partial t^2} \right) \delta \varphi_y
\end{array}
\right] dA
$$

$$(2.177)$$

where the mass moments of inertia can be defined in the following form:

$$
I_i = \int_{-\frac{h}{2}}^{\frac{h}{2}} z^i \rho(z) dz, \quad i = (0,1,2,3,4,6) \tag{2.178}
$$

Now the variation of work done by external loading must be computed to reach the Euler-Lagrange equations of the plate according to the TSDT. The plate is presumed to be subjected to thermal loading. The thermal force applied on the plate is considered to be an in-plane force and it does not possess shear components. Assume the thermal force in x and y directions to be N_x^T and N_y^T, respectively. The variation of work done on the plate by these thermal forces can be formulated as follows:

$$
\delta V = \int_A \left(N_x^T \frac{\partial^2 w}{\partial x^2} + N_y^T \frac{\partial^2 w}{\partial y^2} \right) \delta w \, dA \tag{2.179}
$$

The plate's motion equations can now be easily derived substituting for the variations of strain energy, kinetic energy, and work done by external loading from Eqs. (2.174), (2.177), and (2.179) into the definition of Hamilton's principle, as presented in Eq. (2.21). Once the nontrivial response of the achieved equation is employed, the Euler-Lagrange equations of the plate under thermal loading can be written in the following form:

$$\frac{\partial N_{xx}}{\partial x} + \frac{\partial N_{xy}}{\partial y} - I_0 \frac{\partial^2 u}{\partial t^2} + c_1 I_3 \frac{\partial^3 w}{\partial x \partial t^2} - \left[I_1 - c_1 I_3 \right] \frac{\partial^2 \varphi_x}{\partial t^2} = 0, \qquad (2.180)$$

$$\frac{\partial N_{xy}}{\partial x} + \frac{\partial N_{yy}}{\partial y} - I_0 \frac{\partial^2 v}{\partial t^2} + c_1 I_3 \frac{\partial^3 w}{\partial y \partial t^2} - \left[I_1 - c_1 I_3 \right] \frac{\partial^2 \varphi_y}{\partial t^2} = 0, \qquad (2.181)$$

$$c_1 \left(\frac{\partial^2 P_{xx}}{\partial x^2} + 2 \frac{\partial^2 P_{xy}}{\partial x \partial y} + \frac{\partial^2 P_{yy}}{\partial y^2} \right) + \frac{\partial (Q_{xz} - c_2 S_{xz})}{\partial x}$$

$$+ \frac{\partial (Q_{yz} - c_2 S_{yz})}{\partial y} - \left(N_x^T \frac{\partial^2 w}{\partial x^2} + N_y^T \frac{\partial^2 w}{\partial y^2} \right) - c_1 I_3 \frac{\partial^3 u}{\partial x \partial t^2}$$

$$- c_1 I_3 \frac{\partial^3 v}{\partial y \partial t^2} + I_0 \frac{\partial^2 w}{\partial t^2} + 2 c_1^2 I_6 \frac{\partial^4 w}{\partial x^2 \partial t^2} \qquad (2.182)$$

$$+ c_1 \left[c_1 I_6 - I_4 \right] \frac{\partial^3 \varphi_x}{\partial x \partial t^2} + c_1 \left[c_1 I_6 - I_4 \right] \frac{\partial^3 \varphi_y}{\partial y \partial t^2} = 0,$$

$$\frac{\partial M_{xx}}{\partial x} + \frac{\partial M_{xy}}{\partial y} - c_1 \left(\frac{\partial P_{xx}}{\partial x} + \frac{\partial P_{xy}}{\partial y} \right) + Q_{xz} - c_2 S_{xz} - \left[I_1 - c_1 I_3 \right] \frac{\partial^2 u}{\partial t^2}$$

$$- c_1 \left[c_1 I_6 - I_4 \right] \frac{\partial^3 w}{\partial x \partial t^2} - \left[I_2 - 2 c_1 I_4 + c_1^2 I_6 \right] \frac{\partial^2 \varphi_x}{\partial t^2} = 0, \qquad (2.183)$$

$$\frac{\partial M_{xy}}{\partial x} + \frac{\partial M_{yy}}{\partial y} - c_1 \left(\frac{\partial P_{xy}}{\partial x} + \frac{\partial P_{yy}}{\partial y} \right) + Q_{yz} - c_2 S_{yz} - \left[I_1 - c_1 I_3 \right] \frac{\partial^2 v}{\partial t^2}$$

$$- c_1 \left[c_1 I_6 - I_4 \right] \frac{\partial^3 w}{\partial y \partial t^2} - \left[I_2 - 2 c_1 I_4 + c_1^2 I_6 \right] \frac{\partial^2 \varphi_y}{\partial t^2} = 0 \qquad (2.184)$$

Now the stress-strain relation must be implemented to reach a representation of the previous set of Euler-Lagrange equations free from the stress resultants. Indeed, the equations above must be rewritten in the following steps in terms of the displacement field of the plate. To reach this goal, integrals of the constitutive equation over the thickness of the plate must be computed to be able to replace the stress resultants with their equivalent phrases. Once the mathematical operations are carried out, the following relations can be obtained:

$$
\begin{bmatrix} N_{xx} \\ N_{yy} \\ N_{xy} \end{bmatrix} = \begin{bmatrix} A_{11} & A_{12} & 0 \\ A_{21} & A_{22} & 0 \\ 0 & 0 & A_{66} \end{bmatrix} \begin{bmatrix} \dfrac{\partial u}{\partial x} \\ \dfrac{\partial v}{\partial y} \\ \dfrac{\partial u}{\partial y} + \dfrac{\partial v}{\partial x} \end{bmatrix}
$$

$$
+ \begin{bmatrix} B_{11} - c_1 D_{11} & B_{12} - c_1 D_{12} & 0 \\ B_{21} - c_1 D_{21} & B_{22} - c_1 D_{22} & 0 \\ 0 & 0 & B_{66} - c_1 D_{66} \end{bmatrix} \begin{bmatrix} \dfrac{\partial \varphi_r}{\partial x} \\ \dfrac{\partial \varphi_y}{\partial y} \\ \dfrac{\partial \varphi_x}{\partial y} + \dfrac{\partial \varphi_y}{\partial x} \end{bmatrix} \tag{2.185}
$$

$$
+ \begin{bmatrix} c_1 D_{11} & c_1 D_{12} & 0 \\ c_1 D_{21} & c_1 D_{22} & 0 \\ 0 & 0 & c_1 D_{66} \end{bmatrix} \begin{bmatrix} -\dfrac{\partial^2 w}{\partial x^2} \\ -\dfrac{\partial^2 w}{\partial y^2} \\ -2\dfrac{\partial^2 w}{\partial x \partial y} \end{bmatrix},
$$

$$
\begin{bmatrix} M_{xx} \\ M_{yy} \\ M_{xy} \end{bmatrix} = \begin{bmatrix} B_{11} & B_{12} & 0 \\ B_{21} & B_{22} & 0 \\ 0 & 0 & B_{66} \end{bmatrix} \begin{bmatrix} \dfrac{\partial u}{\partial x} \\ \dfrac{\partial v}{\partial y} \\ \dfrac{\partial u}{\partial y} + \dfrac{\partial v}{\partial x} \end{bmatrix} + \begin{bmatrix} C_{11} - c_1 E_{11} & C_{12} - c_1 E_{12} & 0 \\ C_{21} - c_1 E_{21} & C_{22} - c_1 E_{22} & 0 \\ 0 & 0 & C_{66} - c_1 E_{66} \end{bmatrix} \begin{bmatrix} \dfrac{\partial \varphi_x}{\partial x} \\ \dfrac{\partial \varphi_y}{\partial y} \\ \dfrac{\partial \varphi_x}{\partial y} + \dfrac{\partial \varphi_y}{\partial x} \end{bmatrix}
$$

$$
+ \begin{bmatrix} c_1 E_{11} & c_1 E_{12} & 0 \\ c_1 E_{21} & c_1 E_{22} & 0 \\ 0 & 0 & c_1 E_{66} \end{bmatrix} \begin{bmatrix} -\dfrac{\partial^2 w}{\partial x^2} \\ -\dfrac{\partial^2 w}{\partial y^2} \\ -2\dfrac{\partial^2 w}{\partial x \partial y} \end{bmatrix},
$$

$$
\tag{2.186}
$$

Mechanics of Nanocomposites

$$
\begin{bmatrix} P_{xx} \\ P_{yy} \\ P_{xy} \end{bmatrix} = \begin{bmatrix} D_{11} & D_{12} & 0 \\ D_{21} & D_{22} & 0 \\ 0 & 0 & D_{66} \end{bmatrix} \begin{bmatrix} \dfrac{\partial u}{\partial x} \\ \dfrac{\partial v}{\partial y} \\ \dfrac{\partial u}{\partial y} + \dfrac{\partial v}{\partial x} \end{bmatrix} + \begin{bmatrix} E_{11} - c_1 F_{11} & E_{12} - c_1 F_{12} & 0 \\ E_{21} - c_1 F_{21} & E_{22} - c_1 F_{22} & 0 \\ 0 & 0 & E_{66} - c_1 F_{66} \end{bmatrix} \begin{bmatrix} \dfrac{\partial \varphi_x}{\partial x} \\ \dfrac{\partial \varphi_y}{\partial y} \\ \dfrac{\partial \varphi_x}{\partial y} + \dfrac{\partial \varphi_y}{\partial x} \end{bmatrix}
$$

$$
+ \begin{bmatrix} c_1 F_{11} & c_1 F_{12} & 0 \\ c_1 F_{21} & c_1 F_{22} & 0 \\ 0 & 0 & c_1 F_{66} \end{bmatrix} \begin{bmatrix} -\dfrac{\partial^2 w}{\partial x^2} \\ -\dfrac{\partial^2 w}{\partial y^2} \\ -2\dfrac{\partial^2 w}{\partial x \partial y} \end{bmatrix},
$$

$$
\tag{2.187}
$$

$$
\begin{bmatrix} Q_{xz} \\ Q_{yz} \end{bmatrix} = \begin{bmatrix} G_{55} - c_2 H_{55} & 0 \\ 0 & G_{44} - c_2 H_{44} \end{bmatrix} \begin{bmatrix} \varphi_x + \dfrac{\partial w}{\partial x} \\ \varphi_y + \dfrac{\partial w}{\partial y} \end{bmatrix}, \tag{2.188}
$$

$$
\begin{bmatrix} S_{xz} \\ S_{yz} \end{bmatrix} = \begin{bmatrix} H_{55} - c_2 I_{55} & 0 \\ 0 & H_{44} - c_2 I_{44} \end{bmatrix} \begin{bmatrix} \varphi_x + \dfrac{\partial w}{\partial x} \\ \varphi_y + \dfrac{\partial w}{\partial y} \end{bmatrix} \tag{2.189}
$$

In Eqs. (2.185)–(2.189), the rigidities can be found using the following formula:

$$
\left[A_{ij}, B_{ij}, C_{ij}, D_{ij}, E_{ij}, F_{ij} \right] = \int_{-\frac{h}{2}}^{\frac{h}{2}} \left[1, z, z^2, z^3, z^4, z^6 \right] Q_{ij} dz, \quad i, j = (1, 2, 6),
$$

$$
\tag{2.190}
$$

$$
\left[G_{kl}, H_{kl}, I_{kl} \right] = \int_{-\frac{h}{2}}^{\frac{h}{2}} \left[1, z^2, z^4 \right] Q_{kl} dz, \quad k, l = (4, 5)
$$

The governing equations of the plate can be reached based on the displacement field's components once Eqs. (2.185)–(2.189) are inserted in Eqs. (2.180)–(2.184).

After mathematical operations, the following governing equations can be developed for a third-order plate in the presence of thermal loading:

$$A_{11}\frac{\partial^2 u}{\partial x^2} + A_{12}\frac{\partial^2 v}{\partial x \partial y} + (B_{11} - c_1 D_{11})\frac{\partial^2 \varphi_x}{\partial x^2} + (B_{12} - c_1 D_{12})\frac{\partial^2 \varphi_y}{\partial x \partial y} - c_1 D_{11}\frac{\partial^3 w}{\partial x^3}$$

$$-c_1 D_{12}\frac{\partial^3 w}{\partial x \partial y^2} + A_{66}\left(\frac{\partial^2 u}{\partial y^2} + \frac{\partial^2 v}{\partial x \partial y}\right) + (B_{66} - c_1 D_{66})\left(\frac{\partial^2 \varphi_x}{\partial y^2} + \frac{\partial^2 \varphi_y}{\partial x \partial y}\right)$$ (2.191)

$$-2c_1 D_{66}\frac{\partial^3 w}{\partial x \partial y^2} - I_0\frac{\partial^2 u}{\partial t^2} + c_1 I_3\frac{\partial^3 w}{\partial x \partial t^2} - [I_1 - c_1 I_3]\frac{\partial^2 \varphi_x}{\partial t^2} = 0,$$

$$A_{66}\left(\frac{\partial^2 u}{\partial x \partial y} + \frac{\partial^2 v}{\partial x^2}\right) + (B_{66} - c_1 D_{66})\left(\frac{\partial^2 \varphi_x}{\partial x \partial y} + \frac{\partial^2 \varphi_y}{\partial x^2}\right) - 2c_1 D_{66}\frac{\partial^3 w}{\partial x^2 \partial y}$$

$$+A_{21}\frac{\partial^2 u}{\partial x \partial y} + A_{22}\frac{\partial^2 v}{\partial y^2} + (B_{21} - c_1 D_{21})\frac{\partial^2 \varphi_x}{\partial x \partial y} + (B_{22} - c_1 D_{22})\frac{\partial^2 \varphi_y}{\partial y^2}$$ (2.192)

$$-c_1 D_{21}\frac{\partial^3 w}{\partial x^2 \partial y} - c_1 D_{22}\frac{\partial^3 w}{\partial y^3} - I_0\frac{\partial^2 v}{\partial t^2} + c_1 I_3\frac{\partial^3 w}{\partial y \partial t^2} - [I_1 - c_1 I_3]\frac{\partial^2 \varphi_y}{\partial t^2} = 0,$$

$$c_1\left(\begin{array}{l} D_{11}\dfrac{\partial^3 u}{\partial x^3} + D_{12}\dfrac{\partial^3 v}{\partial x^2 \partial y} + 2D_{66}\left[\dfrac{\partial^3 u}{\partial x \partial y^2} + \dfrac{\partial^3 v}{\partial x^2 \partial y}\right] + D_{21}\dfrac{\partial^3 u}{\partial x \partial y^2} + D_{22}\dfrac{\partial^3 v}{\partial y^3} \\[2mm] +[E_{11} - c_1 F_{11}]\dfrac{\partial^3 \varphi_x}{\partial x^3} + [E_{12} - c_1 F_{12}]\dfrac{\partial^3 \varphi_y}{\partial x^2 \partial y} + 2[E_{66} - c_1 F_{66}]\left[\dfrac{\partial^3 \varphi_x}{\partial x \partial y^2} + \dfrac{\partial^3 \varphi_y}{\partial x^2 \partial y}\right] \\[2mm] +[E_{21} - c_1 F_{21}]\dfrac{\partial^3 \varphi_x}{\partial x \partial y^2} + [E_{22} - c_1 F_{22}]\dfrac{\partial^3 \varphi_y}{\partial y^3} \end{array}\right)$$

$$-c_1^2\left(F_{11}\frac{\partial^4 w}{\partial x^4} + [F_{12} + F_{21} + 2F_{66}]\frac{\partial^4 w}{\partial x^2 \partial y^2} + F_{22}\frac{\partial^4 w}{\partial y^4}\right)$$

$$+(G_{55} - c_2 H_{55} - c_2[H_{55} - c_2 I_{55}])\left(\frac{\partial \varphi_x}{\partial x} + \frac{\partial^2 w}{\partial x^2}\right)$$

$$+(G_{44} - c_2 H_{44} - c_2[H_{44} - c_2 I_{44}])\left(\frac{\partial \varphi_y}{\partial y} + \frac{\partial^2 w}{\partial y^2}\right)$$

$$-\left(N_x^T \frac{\partial^2 w}{\partial x^2} + N_y^T \frac{\partial^2 w}{\partial y^2}\right) - c_1 I_3\frac{\partial^3 u}{\partial x \partial t^2} - c_1 I_3\frac{\partial^3 v}{\partial y \partial t^2}$$

$$+I_0\frac{\partial^2 w}{\partial t^2} + 2c_1^2 I_6\frac{\partial^4 w}{\partial x^2 \partial t^2} + c_1[c_1 I_6 - I_4]\frac{\partial^3 \varphi_x}{\partial x \partial t^2} + c_1[c_1 I_6 - I_4]\frac{\partial^3 \varphi_y}{\partial y \partial t^2} = 0,$$

(2.193)

$$B_{11}\frac{\partial^2 u}{\partial x^2}+B_{12}\frac{\partial^2 v}{\partial x\partial y}+\left(C_{11}-c_1E_{11}\right)\frac{\partial^2 \varphi_x}{\partial x^2}+\left(C_{12}-c_1E_{12}\right)\frac{\partial^2 \varphi_y}{\partial x\partial y}-c_1E_{11}\frac{\partial^3 w}{\partial x^3}-$$

$$c_1E_{12}\frac{\partial^3 w}{\partial x\partial y^2}+B_{66}\left(\frac{\partial^2 u}{\partial y^2}+\frac{\partial^2 v}{\partial x\partial y}\right)+\left(C_{66}-c_1E_{66}\right)\left(\frac{\partial^2 \varphi_x}{\partial y^2}+\frac{\partial^2 \varphi_y}{\partial x\partial y}\right)-2c_1E_{66}\frac{\partial^3 w}{\partial x\partial y^2}-$$

$$c_1\left(\begin{array}{l}D_{11}\dfrac{\partial^2 u}{\partial x^2}+D_{12}\dfrac{\partial^2 v}{\partial x\partial y}+\left[E_{11}-c_1F_{11}\right]\dfrac{\partial^2 \varphi_x}{\partial x^2}+\left[E_{12}-c_1F_{12}\right]\dfrac{\partial^2 \varphi_y}{\partial x\partial y}-c_1F_{11}\dfrac{\partial^3 w}{\partial x^3}-\\[3mm]c_1F_{12}\dfrac{\partial^3 w}{\partial x\partial y^2}+D_{66}\left[\dfrac{\partial^2 u}{\partial y^2}+\dfrac{\partial^2 v}{\partial x\partial y}\right]+\left[E_{66}-c_1F_{66}\right]\left[\dfrac{\partial^2 \varphi_x}{\partial y^2}+\dfrac{\partial^2 \varphi_y}{\partial x\partial y}\right]-2c_1F_{66}\dfrac{\partial^3 w}{\partial x\partial y^2}\end{array}\right)$$

$$+\left(G_{55}-c_2H_{55}\right)\left(\varphi_x+\frac{\partial w}{\partial x}\right)-c_2\left(H_{55}-c_2I_{55}\right)\left(\varphi_x+\frac{\partial w}{\partial x}\right)-\left[I_1-c_1I_3\right]\frac{\partial^2 u}{\partial t^2}-$$

$$c_1\left[c_1I_6-I_4\right]\frac{\partial^3 w}{\partial x\partial t^2}-\left[I_2-2c_1I_4+c_1^2I_6\right]\frac{\partial^2 \varphi_x}{\partial t^2}=0,$$

$$(2.194)$$

$$B_{66}\left(\frac{\partial^2 u}{\partial x\partial y}+\frac{\partial^2 v}{\partial x^2}\right)+\left(C_{66}-c_1E_{66}\right)\left(\frac{\partial^2 \varphi_x}{\partial x\partial y}+\frac{\partial^2 \varphi_y}{\partial x^2}\right)-2c_1E_{66}\frac{\partial^3 w}{\partial x^2\partial y}+B_{21}\frac{\partial^2 u}{\partial x\partial y}$$

$$+B_{22}\frac{\partial^2 v}{\partial y^2}+\left(C_{21}-c_1E_{21}\right)\frac{\partial^2 \varphi_x}{\partial x\partial y}+\left(C_{22}-c_1E_{22}\right)\frac{\partial^2 \varphi_y}{\partial y^2}-c_1E_{21}\frac{\partial^3 w}{\partial x^2\partial y}-c_1E_{22}\frac{\partial^3 w}{\partial y^3}-c_1$$

$$\left(\begin{array}{l}D_{66}\left[\dfrac{\partial^2 u}{\partial x\partial y}+\dfrac{\partial^2 v}{\partial x^2}\right]+\left[E_{66}-c_1F_{66}\right]\left[\dfrac{\partial^2 \varphi_x}{\partial x\partial y}+\dfrac{\partial^2 \varphi_y}{\partial x^2}\right]-2c_1F_{66}\dfrac{\partial^3 w}{\partial x^2\partial y}+D_{21}\dfrac{\partial^2 u}{\partial x\partial y}+\\[3mm]D_{22}\dfrac{\partial^2 v}{\partial y^2}+\left[E_{21}-c_1F_{21}\right]\dfrac{\partial^2 \varphi_x}{\partial x\partial y}+\left[E_{22}-c_1F_{22}\right]\dfrac{\partial^2 \varphi_y}{\partial y^2}-c_1F_{21}\dfrac{\partial^3 w}{\partial x^2\partial y}-c_1F_{22}\dfrac{\partial^3 w}{\partial y^3}\end{array}\right)+$$

$$(G_{44}-c_2H_{44})\left(\varphi_y+\frac{\partial w}{\partial y}\right)-c_2\left(H_{44}-c_2I_{44}\right)\left(\varphi_y+\frac{\partial w}{\partial y}\right)-\left[I_1-c_1I_3\right]\frac{\partial^2 v}{\partial t^2}-$$

$$c_1\left[c_1I_6-I_4\right]\frac{\partial^3 w}{\partial y\partial t^2}-\left[I_2-2c_1I_4+c_1^2I_6\right]\frac{\partial^2 \varphi_y}{\partial t^2}=0$$

$$(2.195)$$

The governing equations have been obtained and both natural frequency and critical buckling load of the plate can be reached solving the equations above for dynamic and static cases, respectively. For the goal of solving the equations above, Navier's type analytical solution will be selected to solve the equations as well as to satisfy the simply supported BC at all of the edges. The solution functions for the components of the displacement field can be assumed to be in the following form on the basis of the Navier's method:

$$u = \sum_{m=1}^{\infty}\sum_{n=1}^{\infty} U_{mn} \cos\left(\frac{m\pi x}{a}\right)\sin\left(\frac{n\pi y}{b}\right)e^{i\omega t},$$

$$v = \sum_{m=1}^{\infty}\sum_{n=1}^{\infty} V_{mn} \sin\left(\frac{m\pi x}{a}\right)\cos\left(\frac{n\pi y}{b}\right)e^{i\omega t},$$

$$w = \sum_{m=1}^{\infty}\sum_{n=1}^{\infty} W_{mn} \sin\left(\frac{m\pi x}{a}\right)\sin\left(\frac{n\pi y}{b}\right)e^{i\omega t},$$

$$\varphi_x = \sum_{m=1}^{\infty}\sum_{n=1}^{\infty} \Phi_{xmn} \cos\left(\frac{m\pi x}{a}\right)\sin\left(\frac{n\pi y}{b}\right)e^{i\omega t},$$

$$\varphi_y = \sum_{m=1}^{\infty}\sum_{n=1}^{\infty} \Phi_{ymn} \sin\left(\frac{m\pi x}{a}\right)\cos\left(\frac{n\pi y}{b}\right)e^{i\omega t}$$

(2.196)

where U_{mn}, V_{mn}, W_{mn}, Φ_{xmn}, and Φ_{ymn} are the amplitudes for the plate's motion. Moreover, m and n are mode numbers in x and y directions, respectively. In the dynamic problems, ω reveals the natural frequency of the plate, whereas this term will disappear in static buckling analyses. It is noteworthy that substitution of Eq. (2.196) in Eqs. (2.191)–(2.195) results in reaching a 5×5 eigenvalue equation. The natural frequency of the plate will be attained by solving the obtained equation for ω. Once the buckling response is extracted, a buckling load must be added to the governing equations the same as a thermal force and afterward, the critical value of this buckling load will be reached by setting the determinant of the stiffness matrix to be equal with zero and solving the problem for the buckling load.

2.2.7 REFINED HIGHER-ORDER PLATE THEORY

This section presents mathematical formulations for the vibration problem of a plate in the framework of a refined higher-order plate hypothesis. Based on refined theories, the impacts of shear stress and strain on the motion equations will be included without employing shear correction factors with four variables. Indeed, the most crucial superiority of the refined theory of plates compared with the simple higher-order ones is in the lower computational cost needed to reach the exact response. Once the dynamic form of the governing equations are obtained, the motion equations for the static buckling problem can be developed by eliminating the time-dependent terms. Now, we must begin with introduction of the displacement field of such theories. According to refined plate models, the displacement field can be written as follows:

$$u_x(x,y,z,t) = u(x,y,t) - z\frac{\partial w_b(x,y,t)}{\partial x} - f(z)\frac{\partial w_s(x,y,t)}{\partial x},$$

$$u_y(x,y,z,t) = v(x,y,t) - z\frac{\partial w_b(x,y,t)}{\partial y} - f(z)\frac{\partial w_s(x,y,t)}{\partial y}, \qquad (2.197)$$

$$u_z(x,y,z,t) = w_b(x,y,t) + w_s(x,y,t)$$

where u and v are the longitudinal and transverse displacements, respectively. Also, the total deflection of the plate is divided into two bending (w_b) and shear (w_s) deflections. The function $f(z)$ is presented in the previous formulas to satisfy the shear free condition at the upper and lower surfaces and to govern the distribution of the shear stress and strain across the thickness of the plate. Among the various types of shape functions that are widely used by researchers, in this book, the following shape function will be utilized in Eq. (2.197):

$$f(z) = z - \frac{4z^3}{3h^2} \qquad (2.198)$$

The components of the strain tensor must now be calculated. To this purpose, the displacement field of refined higher-order plates, presented in Eq. (2.197), must be combined with the definition of the Eulerian strain tensor, introduced in Eq. (2.18). Once the mathematical combination is completed; the nonzero strains of refined plates can be attained as follows:

$$\varepsilon_{xx} = \frac{1}{2}\left(\frac{\partial u_x}{\partial x} + \frac{\partial u_x}{\partial x}\right) = \frac{\partial u}{\partial x} - z\frac{\partial^2 w_b}{\partial x^2} - f(z)\frac{\partial^2 w_s}{\partial x^2},$$

$$\varepsilon_{yy} = \frac{1}{2}\left(\frac{\partial u_y}{\partial y} + \frac{\partial u_y}{\partial y}\right) = \frac{\partial v}{\partial y} - z\frac{\partial^2 w_b}{\partial y^2} - f(z)\frac{\partial^2 w_s}{\partial y^2},$$

$$\gamma_{xy} = 2\varepsilon_{xy} = \left(\frac{\partial u_x}{\partial y} + \frac{\partial u_y}{\partial x}\right) = \frac{\partial u}{\partial y} + \frac{\partial v}{\partial x} - 2z\frac{\partial^2 w_b}{\partial x\partial y} - 2f(z)\frac{\partial^2 w_s}{\partial x\partial y},$$

$$\gamma_{xz} = 2\varepsilon_{xz} = \left(\frac{\partial u_x}{\partial z} + \frac{\partial u_z}{\partial x}\right) = g(z)\frac{\partial w_s}{\partial x},$$

$$\gamma_{yz} = 2\varepsilon_{yz} = \left(\frac{\partial u_y}{\partial z} + \frac{\partial u_z}{\partial y}\right) = g(z)\frac{\partial w_s}{\partial y}$$

$$(2.199)$$

where

$$g(z) = 1 - \frac{df(z)}{dz} \tag{2.200}$$

Now the energy-based variations must be calculated to enrich the Euler-Lagrange equations of refined shear deformable plates. First the variation of strain energy must be calculated. The definition of this term is as same as that presented in Eq. (2.173). Mixing the strain-displacement relationships of refined plates (see Eq. (2.199)) with Eq. (2.173) reveals the following expression for the variation of the strain energy of the plate:

$$\delta U = \int_A \left[\begin{array}{l} N_{xx}\left(\delta\frac{\partial u}{\partial x}\right) - M_{xx}^b\left(\delta\frac{\partial^2 w_b}{\partial x^2}\right) - M_{xx}^s\left(\delta\frac{\partial^2 w_s}{\partial x^2}\right) + N_{yy}\left(\delta\frac{\partial v}{\partial y}\right) - \\[3mm] M_{yy}^b\left(\delta\frac{\partial^2 w_b}{\partial y^2}\right) - M_{yy}^s\left(\delta\frac{\partial^2 w_s}{\partial y^2}\right) + N_{xy}\left(\delta\left[\frac{\partial u}{\partial y}+\frac{\partial v}{\partial x}\right]\right) - M_{xy}^b\left(2\delta\frac{\partial^2 w_b}{\partial x \partial y}\right) \\[3mm] - M_{xy}^s\left(2\delta\frac{\partial^2 w_s}{\partial x \partial y}\right) + Q_{xz}\left(\delta\frac{\partial w_s}{\partial x}\right) + Q_{yz}\left(\delta\frac{\partial w_s}{\partial y}\right) \end{array} \right] dA$$

$$\tag{2.201}$$

In the equation above, the stress resultants can be defined as follows:

$$\left[N_{xx}, N_{yy}, N_{xy} \right] = \int_{-\frac{h}{2}}^{\frac{h}{2}} \left[\sigma_{xx}, \sigma_{yy}, \sigma_{xy} \right] dz,$$

$$\left[M_{xx}^b, M_{yy}^b, M_{xy}^b \right] = \int_{-\frac{h}{2}}^{\frac{h}{2}} z\left[\sigma_{xx}, \sigma_{yy}, \sigma_{xy} \right] dz,$$

$$\tag{2.202}$$

$$\left[M_{xx}^s, M_{yy}^s, M_{xy}^s \right] = \int_{-\frac{h}{2}}^{\frac{h}{2}} f(z)\left[\sigma_{xx}, \sigma_{yy}, \sigma_{xy} \right] dz,$$

$$\left[Q_{xz}, Q_{yz} \right] = \int_{-\frac{h}{2}}^{\frac{h}{2}} g(z)\left[\sigma_{xz}, \sigma_{yz} \right] dz$$

In addition, the variation of the kinetic energy must be computed. The definition of this term was previously mentioned in Eq. (2.176). Inserting Eq. (2.197) in Eq. (2.176) results in the following expression for the variation of the kinetic energy:

$$
\delta T = \int\limits_{A}
\begin{bmatrix}
I_0\left(\dfrac{\partial u}{\partial t}\dfrac{\partial \delta u}{\partial t} + \dfrac{\partial v}{\partial t}\dfrac{\partial \delta v}{\partial t} + \dfrac{\partial (w_b + w_s)}{\partial t}\dfrac{\partial \delta(w_b + w_s)}{\partial t}\right) \\[2ex]
-I_1\left(\dfrac{\partial u}{\partial t}\dfrac{\partial^2 \delta w_b}{\partial x \partial t} + \dfrac{\partial \delta u}{\partial t}\dfrac{\partial^2 w_b}{\partial x \partial t} + \dfrac{\partial v}{\partial t}\dfrac{\partial^2 \delta w_b}{\partial y \partial t} + \dfrac{\partial \delta v}{\partial t}\dfrac{\partial^2 w_b}{\partial y \partial t}\right) \\[2ex]
-J_1\left(\dfrac{\partial u}{\partial t}\dfrac{\partial^2 \delta w_s}{\partial x \partial t} + \dfrac{\partial \delta u}{\partial t}\dfrac{\partial^2 w_s}{\partial x \partial t} + \dfrac{\partial v}{\partial t}\dfrac{\partial^2 \delta w_s}{\partial y \partial t} + \dfrac{\partial \delta v}{\partial t}\dfrac{\partial^2 w_s}{\partial y \partial t}\right) \\[2ex]
+I_2\left(\dfrac{\partial^2 w_b}{\partial x \partial t}\dfrac{\partial^2 \delta w_b}{\partial x \partial t} + \dfrac{\partial^2 w_b}{\partial y \partial t}\dfrac{\partial^2 \delta w_b}{\partial y \partial t}\right) \\[2ex]
+K_2\left(\dfrac{\partial^2 w_s}{\partial x \partial t}\dfrac{\partial^2 \delta w_s}{\partial x \partial t} + \dfrac{\partial^2 w_s}{\partial y \partial t}\dfrac{\partial^2 \delta w_s}{\partial y \partial t}\right) \\[2ex]
+J_2\left(\dfrac{\partial^2 w_b}{\partial x \partial t}\dfrac{\partial^2 \delta w_s}{\partial x \partial t} + \dfrac{\partial^2 w_s}{\partial x \partial t}\dfrac{\partial^2 \delta w_b}{\partial x \partial t} + \dfrac{\partial^2 w_b}{\partial y \partial t}\dfrac{\partial^2 \delta w_s}{\partial y \partial t} + \dfrac{\partial^2 w_s}{\partial y \partial t}\dfrac{\partial^2 \delta w_b}{\partial y \partial t}\right)
\end{bmatrix} dA \quad (2.203)
$$

where the mass moments of inertia utilized in the equation above can be defined in the following form:

$$
[I_0, I_1, J_1, I_2, J_2, K_2] = \int\limits_{-\frac{h}{2}}^{\frac{h}{2}} \left[1, z, f(z), z^2, zf(z), f^2(z)\right]\rho(z)dz \quad (2.204)
$$

Now it is time to generate an expression for the variation of the work done by external forces. The structure is assumed to be embedded on a three-variable visco-Pasternak medium and placed in an environment with thermal gradient. The elastic coefficients of the medium's springs will be shown with signs k_w and k_p for Winkler and Pasternak coefficients, respectively. The damping coefficient of the viscoelastic foundation will be shown with c_d. Also, the thermal loading applied on the plate will be shown with N^T. Now, the variation of the work done by external forces can be presented in the following form:

$$
\delta V = \int\limits_{A}\left[k_w(w_b + w_s) + (k_p - N^T)\left(\frac{\partial^2(w_b + w_s)}{\partial x^2} + \frac{\partial^2(w_b + w_s)}{\partial y^2}\right) - c_d\frac{\partial(w_b + w_s)}{\partial t}\right]\delta(w_b + w_s)dA
$$

$$(2.205)$$

The Euler-Lagrange equations can be reached once the variations of strain energy, kinetic energy, and work done by external forces are substituted from Eqs. (2.201), (2.203), and (2.205) into Eq. (2.21). After the aforementioned substitution, the expressions multiplied by the variations of the components of the displacement field of the plate must be considered to be identical with zero. Finally, the Euler-Lagrange equations of the plate under defined loading condition can be reached in the following form:

$$\frac{\partial N_{xx}}{\partial x} + \frac{\partial N_{xy}}{\partial y} = I_0 \frac{\partial^2 u}{\partial t^2} - I_1 \frac{\partial^3 w_b}{\partial x \partial t^2} - J_1 \frac{\partial^3 w_s}{\partial x \partial t^2}, \tag{2.206}$$

$$\frac{\partial N_{xy}}{\partial x} + \frac{\partial N_{yy}}{\partial y} = I_0 \frac{\partial^2 v}{\partial t^2} - I_1 \frac{\partial^3 w_b}{\partial y \partial t^2} - J_1 \frac{\partial^3 w_s}{\partial y \partial t^2}, \tag{2.207}$$

$$\frac{\partial^2 M_{xx}^b}{\partial x^2} + 2\frac{\partial^2 M_{xy}^b}{\partial x \partial y} + \frac{\partial^2 M_{yy}^b}{\partial y^2} - k_w(w_b + w_s)$$

$$+ (k_p - N^T)\left(\frac{\partial^2(w_b + w_s)}{\partial x^2} + \frac{\partial^2(w_b + w_s)}{\partial y^2}\right) - c_d$$

$$\frac{\partial(w_b + w_s)}{\partial t} = I_0 \frac{\partial^2(w_b + w_s)}{\partial t^2} + I_1\left(\frac{\partial^3 u}{\partial x \partial t^2} + \frac{\partial^3 v}{\partial y \partial t^2}\right) \tag{2.208}$$

$$- I_2\left(\frac{\partial^4 w_b}{\partial x^2 \partial t^2} + \frac{\partial^4 w_b}{\partial y^2 \partial t^2}\right) - J_2\left(\frac{\partial^4 w_s}{\partial x^2 \partial t^2} + \frac{\partial^4 w_s}{\partial y^2 \partial t^2}\right),$$

$$\frac{\partial^2 M_{xx}^s}{\partial x^2} + 2\frac{\partial^2 M_{xy}^s}{\partial x \partial y} + \frac{\partial^2 M_{yy}^s}{\partial y^2} + \frac{\partial Q_{xz}}{\partial x} + \frac{\partial Q_{yz}}{\partial y} - k_w(w_b + w_s)$$

$$+ (k_p - N^T)\left(\frac{\partial^2(w_b + w_s)}{\partial x^2} + \frac{\partial^2(w_b + w_s)}{\partial y^2}\right)$$

$$- c_d\frac{\partial(w_b + w_s)}{\partial t} = I_0 \frac{\partial^2(w_b + w_s)}{\partial t^2} + J_1\left(\frac{\partial^3 u}{\partial x \partial t^2} + \frac{\partial^3 v}{\partial y \partial t^2}\right) \tag{2.209}$$

$$- J_2\left(\frac{\partial^4 w_b}{\partial x^2 \partial t^2} + \frac{\partial^4 w_b}{\partial y^2 \partial t^2}\right) - K_2\left(\frac{\partial^4 w_s}{\partial x^2 \partial t^2} + \frac{\partial^4 w_s}{\partial y^2 \partial t^2}\right)$$

The stress-strain relations must be used to derive equivalent equations for the stress resultants with the goal of presenting the equations above in terms of the displacement field of the plate. Some integrals must be calculated over the thickness of the plate. In all of these integrals, the arrays of the plate's stiffness matrix will be involved. After mathematical simplifications, the stress resultants can be presented in the following form:

$$
\begin{bmatrix} N_{xx} \\ N_{yy} \\ N_{xy} \end{bmatrix} = \begin{bmatrix} A_{11} & A_{12} & 0 \\ A_{21} & A_{22} & 0 \\ 0 & 0 & A_{66} \end{bmatrix} \begin{bmatrix} \dfrac{\partial u}{\partial x} \\ \dfrac{\partial v}{\partial y} \\ \dfrac{\partial u}{\partial y} + \dfrac{\partial v}{\partial x} \end{bmatrix}
$$

$$
+ \begin{bmatrix} B_{11} & B_{12} & 0 \\ B_{21} & B_{22} & 0 \\ 0 & 0 & B_{66} \end{bmatrix} \begin{bmatrix} -\dfrac{\partial^2 w_b}{\partial x^2} \\ -\dfrac{\partial^2 w_b}{\partial y^2} \\ -2\dfrac{\partial^2 w_b}{\partial x \partial y} \end{bmatrix} \tag{2.210}
$$

$$
+ \begin{bmatrix} B_{11}^s & B_{12}^s & 0 \\ B_{21}^s & B_{22}^s & 0 \\ 0 & 0 & B_{66}^s \end{bmatrix} \begin{bmatrix} -\dfrac{\partial^2 w_s}{\partial x^2} \\ -\dfrac{\partial^2 w_s}{\partial y^2} \\ -2\dfrac{\partial^2 w_s}{\partial x \partial y} \end{bmatrix},
$$

$$
\begin{bmatrix} M_{xx}^b \\ M_{yy}^b \\ M_{xy}^b \end{bmatrix} = \begin{bmatrix} B_{11} & B_{12} & 0 \\ B_{21} & B_{22} & 0 \\ 0 & 0 & B_{66} \end{bmatrix} \begin{bmatrix} \dfrac{\partial u}{\partial x} \\ \dfrac{\partial v}{\partial y} \\ \dfrac{\partial u}{\partial y} + \dfrac{\partial v}{\partial x} \end{bmatrix} + \begin{bmatrix} D_{11} & D_{12} & 0 \\ D_{21} & D_{22} & 0 \\ 0 & 0 & D_{66} \end{bmatrix} \begin{bmatrix} -\dfrac{\partial^2 w_b}{\partial x^2} \\ -\dfrac{\partial^2 w_b}{\partial y^2} \\ -2\dfrac{\partial^2 w_b}{\partial x \partial y} \end{bmatrix}
$$

$$
+ \begin{bmatrix} D_{11}^s & D_{12}^s & 0 \\ D_{21}^s & D_{22}^s & 0 \\ 0 & 0 & D_{66}^s \end{bmatrix} \begin{bmatrix} -\dfrac{\partial^2 w_s}{\partial x^2} \\ -\dfrac{\partial^2 w_s}{\partial y^2} \\ -2\dfrac{\partial^2 w_s}{\partial x \partial y} \end{bmatrix}, \tag{2.211}
$$

$$
\begin{bmatrix} M_{xx}^s \\ M_{yy}^s \\ M_{xy}^s \end{bmatrix} = \begin{bmatrix} B_{11}^s & B_{12}^s & 0 \\ B_{21}^s & B_{22}^s & 0 \\ 0 & 0 & B_{66}^s \end{bmatrix} \begin{bmatrix} \dfrac{\partial u}{\partial x} \\[4pt] \dfrac{\partial v}{\partial y} \\[4pt] \dfrac{\partial u}{\partial y} + \dfrac{\partial v}{\partial x} \end{bmatrix} + \begin{bmatrix} D_{11}^s & D_{12}^s & 0 \\ D_{21}^s & D_{22}^s & 0 \\ 0 & 0 & D_{66}^s \end{bmatrix} \begin{bmatrix} -\dfrac{\partial^2 w_b}{\partial x^2} \\[4pt] -\dfrac{\partial^2 w_b}{\partial y^2} \\[4pt] -2\dfrac{\partial^2 w_b}{\partial x \partial y} \end{bmatrix}
$$

$$
+ \begin{bmatrix} H_{11}^s & H_{12}^s & 0 \\ H_{21}^s & H_{22}^s & 0 \\ 0 & 0 & H_{66}^s \end{bmatrix} \begin{bmatrix} -\dfrac{\partial^2 w_s}{\partial x^2} \\[4pt] -\dfrac{\partial^2 w_s}{\partial y^2} \\[4pt] -2\dfrac{\partial^2 w_s}{\partial x \partial y} \end{bmatrix},
$$

(2.212)

$$
\begin{bmatrix} Q_{xz} \\ Q_{yz} \end{bmatrix} = \begin{bmatrix} A_{55}^s & 0 \\ 0 & A_{44}^s \end{bmatrix} \begin{bmatrix} \dfrac{\partial w_s}{\partial x} \\[4pt] \dfrac{\partial w_s}{\partial y} \end{bmatrix}
$$

(2.213)

In Eqs. (2.210)–(2.213), the rigidities can be defined in the following form:

$$
\left[A_{ij}, B_{ij}, B_{ij}^s, D_{ij}, D_{ij}^s, H_{ij}^s \right] = \int_{-\frac{h}{2}}^{\frac{h}{2}} \left[1, z, f(z), z^2, zf(z), f^2(z) \right] Q_{ij} dz, \quad i,j = (1,2,6)
$$

(2.214)

$$
\left[A_{44}^s, A_{55}^s \right] = \int_{-\frac{h}{2}}^{\frac{h}{2}} \left[Q_{44}, Q_{55} \right] g^2(z) dz
$$

Now the governing equations of motion can be derived in terms of the displacement field of the plate by inserting Eqs. (2.210)–(2.213) in Eqs. (2.206)–(2.209). Once this substitution is accomplished, the governing equations can be presented in the following form:

$$
A_{11} \frac{\partial^2 u}{\partial x^2} + A_{66} \frac{\partial^2 u}{\partial y^2} + (A_{12} + A_{66}) \frac{\partial^2 v}{\partial x \partial y} - B_{11} \frac{\partial^3 w_b}{\partial x^3} - (B_{12} + 2B_{66}) \frac{\partial^3 w_b}{\partial x \partial y^2}
$$
$$
- B_{11}^s \frac{\partial^3 w_s}{\partial x^3} - (B_{12}^s + 2B_{66}^s) \frac{\partial^3 w_s}{\partial x \partial y^2} - I_0 \frac{\partial^2 u}{\partial t^2} + I_1 \frac{\partial^3 w_b}{\partial x \partial t^2} + J_1 \frac{\partial^3 w_s}{\partial x \partial t^2} = 0,
$$

(2.215)

$$
\left(A_{12} + A_{66} \right) \frac{\partial^2 u}{\partial x \partial y} + A_{66} \frac{\partial^2 v}{\partial x^2} + A_{22} \frac{\partial^2 v}{\partial y^2} - \left(B_{12} + 2B_{66} \right) \frac{\partial^3 w_b}{\partial x^2 \partial y} - B_{22} \frac{\partial^3 w_b}{\partial y^3}
$$

$$
- \left(B_{12}^s + 2B_{66}^s \right) \frac{\partial^3 w_s}{\partial x^2 \partial y} - B_{22}^s \frac{\partial^3 w_s}{\partial y^3} - I_0 \frac{\partial^2 v}{\partial t^2} + I_1 \frac{\partial^3 w_b}{\partial y \partial t^2} + J_1 \frac{\partial^3 w_s}{\partial y \partial t^2} = 0
$$

(2.216)

$$
B_{11} \frac{\partial^3 u}{\partial x^3} + \left(B_{12} + 2B_{66} \right) \left(\frac{\partial^3 u}{\partial x \partial y^2} + \frac{\partial^3 v}{\partial x^2 \partial y} \right) + B_{22} \frac{\partial^3 v}{\partial y^3} - D_{11} \frac{\partial^4 w_b}{\partial x^4}
$$

$$
-2 \left(D_{12} + 2D_{66} \right) \frac{\partial^4 w_b}{\partial x^2 \partial y^2} - D_{22} \frac{\partial^4 w_b}{\partial y^4}
$$

$$
- D_{11}^s \frac{\partial^4 w_s}{\partial x^4} - 2 \left(D_{12}^s + 2D_{66}^s \right) \frac{\partial^4 w_s}{\partial x^2 \partial y^2} - D_{22}^s \frac{\partial^4 w_s}{\partial y^4} - k_w \left(w_b + w_s \right)
$$

(2.217)

$$
+ \left(k_p - N^T \right) \left(\frac{\partial^2 \left(w_b + w_s \right)}{\partial x^2} + \frac{\partial^2 \left(w_b + w_s \right)}{\partial y^2} \right)
$$

$$
- c_d \frac{\partial \left(w_b + w_s \right)}{\partial t} - I_0 \frac{\partial^2 \left(w_b + w_s \right)}{\partial t^2} - I_1 \left(\frac{\partial^3 u}{\partial x \partial t^2} + \frac{\partial^3 v}{\partial y \partial t^2} \right)
$$

$$
+ I_2 \left(\frac{\partial^4 w_b}{\partial x^2 \partial t^2} + \frac{\partial^4 w_b}{\partial y^2 \partial t^2} \right) + J_2 \left(\frac{\partial^4 w_s}{\partial x^2 \partial t^2} + \frac{\partial^4 w_s}{\partial y^2 \partial t^2} \right) = 0
$$

$$
B_{11}^s \frac{\partial^3 u}{\partial x^3} + \left(B_{12}^s + 2B_{66}^s \right) \left(\frac{\partial^3 u}{\partial x \partial y^2} + \frac{\partial^3 v}{\partial x^2 \partial y} \right) + B_{22}^s \frac{\partial^3 v}{\partial y^3} - D_{11}^s \frac{\partial^4 w_b}{\partial x^4}
$$

$$
-2 \left(D_{12}^s + 2D_{66}^s \right) \frac{\partial^4 w_b}{\partial x^2 \partial y^2} - D_{22}^s \frac{\partial^4 w_b}{\partial y^4} - H_{11}^s \frac{\partial^4 w_s}{\partial x^4} - 2 \left(H_{12}^s + 2H_{66}^s \right) \frac{\partial^4 w_s}{\partial x^2 \partial y^2}
$$

$$
- H_{22}^s \frac{\partial^4 w_s}{\partial y^4} - k_w \left(w_b + w_s \right) + \left(k_p - N^T \right) \left(\frac{\partial^2 \left(w_b + w_s \right)}{\partial x^2} + \frac{\partial^2 \left(w_b + w_s \right)}{\partial y^2} \right)
$$

(2.218)

$$
- c_d \frac{\partial \left(w_b + w_s \right)}{\partial t} - I_0 \frac{\partial^2 \left(w_b + w_s \right)}{\partial t^2} - J_1 \left(\frac{\partial^3 u}{\partial x \partial t^2} + \frac{\partial^3 v}{\partial y \partial t^2} \right)
$$

$$
+ J_2 \left(\frac{\partial^4 w_b}{\partial x^2 \partial t^2} + \frac{\partial^4 w_b}{\partial y^2 \partial t^2} \right) + K_2 \left(\frac{\partial^4 w_s}{\partial x^2 \partial t^2} + \frac{\partial^4 w_s}{\partial y^2 \partial t^2} \right) = 0
$$

Now the governing equations of the dynamic problem are in hand. It is worth mentioning that the governing equations of the static buckling problem can be reached whenever terms with differentiations with respect to time are ignored and a buckling load parameter is added to the governing equations in the same manner as thermal force. The static and dynamic responses of the problem are now addressed. In this section, the well-known Navier's solution will be used to generate the mechanical response of the problem. Based on this method, the simply supported BC at all of the plate's edges will be satisfied. The displacement field of the plate will possess the following solution functions based upon this method:

$$
u = \sum_{m=1}^{\infty} \sum_{n=1}^{\infty} U_{mn} \cos\left(\frac{m\pi x}{a}\right) \sin\left(\frac{n\pi y}{b}\right) e^{i\omega t},
$$

$$
v = \sum_{m=1}^{\infty} \sum_{n=1}^{\infty} V_{mn} \sin\left(\frac{m\pi x}{a}\right) \cos\left(\frac{n\pi y}{b}\right) e^{i\omega t},
$$

$$
w_b = \sum_{m=1}^{\infty} \sum_{n=1}^{\infty} W_{bmn} \sin\left(\frac{m\pi x}{a}\right) \sin\left(\frac{n\pi y}{b}\right) e^{i\omega t},
$$

$$
w_s = \sum_{m=1}^{\infty} \sum_{n=1}^{\infty} W_{smn} \sin\left(\frac{m\pi x}{a}\right) \sin\left(\frac{n\pi y}{b}\right) e^{i\omega t}
$$

(2.219)

In the equations above, U_{mn}, V_{mn}, W_{bmn}, and W_{smn} are the unknown amplitudes. Also, the natural frequency is ω. Similar to the previous procedures explained for solving the governing equations, Eq. (2.219) must be inserted in Eqs. (2.215)–(2.218). Once the substitution is carried out, an eigenvalue equation will be developed that results in extracting the natural frequency of the plate. In addition, the static buckling problem corresponding with Eqs. (2.215)–(2.218) can be solved by eliminating the mass and damping matrices of the vibration problem and only considering the terms of the stiffness matrix.

Another dynamic response that may be explored is the transient response of the system. In fact, this response belongs to the variation of the bending deflection of the system versus time. To reach this answer, the conventional methods of extracting the natural frequency cannot be implemented. Indeed, such a problem can be solved by applying a transformation to change the problem and move it from the temporal domain to another one followed by an inversion. In fact, the time-dependent (transient) response of the problem can be achieved using one of the well-known transformations to solve the problem in the transformed area and reach the mechanical response by inversing the solved problem to the primary area (i.e., time in such problems). In this book, one of the best methods of transformation will be implemented to reach the transient response of a refined plate. This method is the Laplace transformation technique and the instructions about how to apply it will be explained in the following steps. For the sake of completeness, the general form of an oscillating system will be discussed here. Consider the following eigenvalue equation for the dynamic problem of a refined higher-order plate:

$$
\left[K + C\frac{\partial}{\partial t} + M\frac{\partial^2}{\partial t^2} \right]\{\Delta\} = \{F\}
$$

(2.220)

where **K**, **M**, and **C** are stiffness, mass, and damping matrices, respectively. Also, vectors Δ and **F** denote the displacement field and loading vectors, respectively. To reach the transient response of such a problem, the solution function of the displacement field's components must be presented in the following form:

$$u = \sum_{m=1}^{\infty}\sum_{n=1}^{\infty} U_{mn}(t)\cos\left(\frac{m\pi x}{a}\right)\sin\left(\frac{n\pi y}{b}\right),$$

$$v = \sum_{m=1}^{\infty}\sum_{n=1}^{\infty} V_{mn}(t)\sin\left(\frac{m\pi x}{a}\right)\cos\left(\frac{n\pi y}{b}\right),$$

$$w_b = \sum_{m=1}^{\infty}\sum_{n=1}^{\infty} W_{bmn}(t)\sin\left(\frac{m\pi x}{a}\right)\sin\left(\frac{n\pi y}{b}\right),$$

$$w_s = \sum_{m=1}^{\infty}\sum_{n=1}^{\infty} W_{smn}(t)\sin\left(\frac{m\pi x}{a}\right)\sin\left(\frac{n\pi y}{b}\right)$$

(2.221)

Also, the loading vector for the case of a dynamic bending problem can be presented in the following form:

$$\{\mathbf{F}\} = \begin{Bmatrix} 0 \\ 0 \\ q(x,y,t) \\ q(x,y,t) \end{Bmatrix}$$

(2.222)

in which

$$q(x,y,t) = \sum_{m=1}^{\infty}\sum_{n=1}^{\infty} q_{mn}(t)\sin\left(\frac{m\pi x}{a}\right)\sin\left(\frac{n\pi y}{b}\right)$$

(2.223)

where $q_{mn}(t)$ is the time-dependent part of the transverse loading once a dynamic analysis is performed. Herein, a uniform loading is applied to the plate and the $q_{mn}(t)$ can be expressed in the following form:

$$q_{mn}(t) = \frac{16Q(t)}{mn\pi^2}$$

(2.224)

where m and n are only allowed to be odd integers. Herein, the term $Q(t)$ can be presented in the following form:

$$Q(t) = q_0 H(t)$$

(2.225)

in which $H(t)$ is the Heaviside step function that can be mathematically presented as follows:

$$H(t) = \begin{cases} 0 & t < 0 \\ 1 & t \geq 0 \end{cases}$$

(2.226)

Eq. (2.221) must now be substituted in Eqs. (2.215)–(2.218) and the effect of the applied bending loading must be considered to reach the final form of the problem as follows:

$$[\mathbf{K}]\{\Delta_{mn}\}+[\mathbf{C}]\{\dot{\Delta}_{mn}\}+[\mathbf{M}]\{\ddot{\Delta}_{mn}\}=\{\mathbf{F}\} \qquad (2.227)$$

where $\Delta_{\mathbf{mn}}$ is a vector containing the time-dependent part of the displacements and can be considered as $\Delta_{\mathbf{mn}}=\left\{\ U_{mn}\ V_{mn}\ W_{bmn}\ W_{smn}\ \right\}^{T}$. To solve this problem in the time domain, a transformation is required, followed by an inversion. It is very useful to employ the Laplace transformation to map the final eigenvalue equation from time domain to frequency or s- domain and solve the problem for the displacements in the frequency domain and afterwards return to the time domain via an analytic Laplace inversion. Prior to solving the present problem, it can be useful to review the fundamentals of Laplace transformation. Assume a function $f(t)$ in the time domain and its mapped component in the s- domain as $F(s)$. These functions can be related together via:

$$\mathcal{L}\{f(t)\}=F(s)=\int_{0}^{\infty}e^{-st}f(t)dt \qquad (2.228)$$

After implementing Eq. (2.228) and applying it to Eq. (2.227), the transformed eigenvalue equation of the problem can be presented in the following form in the transformed Laplace domain:

$$\left([\mathbf{K}]+s[\mathbf{C}]+s^{2}[\mathbf{M}]\right)\{\overline{\Delta}_{mn}\}=\{\overline{F}\} \qquad (2.229)$$

In the equation above, the bar sign stands for the transformed variants in the frequency domain. Now it is simple to find the displacement vector in the frequency domain via following mathematical relation:

$$\{\overline{\Delta}_{mn}\}=\left([\mathbf{K}]+s[\mathbf{C}]+s^{2}[\mathbf{M}]\right)^{-1}\{\overline{F}\} \qquad (2.230)$$

The transformed displacements can be determined according to the equation above. The next step is finding the displacements in the time domain. The solution can be completed by calculating the inverse Laplace transformation of the Eq. (2.230). However, such a mathematical operation cannot be applied as simply as that of Laplace transformation. Indeed, there is a difference between the Laplace transformation and its inversion and that difference relates to the issue of the region of convergence (ROC), which must be deeply considered when an inverse Laplace transformation is computed. Following the well-known Bromwich integral, the inverse Laplace transformation of a desirable function $F(s)$ can be calculated in the following form:

$$\mathcal{L}^{-1}\{F(s)\}=f(t)=\frac{1}{2\pi i}\int_{c-i\infty}^{c+i\infty}e^{st}F(s)ds \qquad (2.231)$$

The significance of ROC can be inferred from the previous relation. Actually, the previous integral must be calculated along the path from $c - i\infty$ to $c + i\infty$ for any c which is in the ROC. In the following steps, the issue of calculating the inverse Laplace transformation will be extended to reach the problem's response.

Consider the Laplace transformation of a desirable function $F(s)$ to be in the following form:

$$F(s) = \frac{N(s)}{D(s)} = \frac{a_k s^k + a_{k-1} s^{k-1} + ... + a_1 s + a_0}{b_l s^k + b_{l-1} s^{k-1} + ... + b_1 s + b_0} \qquad (2.232)$$

As stated in Eq. (2.232), the $N(s)$ and $D(s)$ functions are polynomials and they do not possess common zeros. Assume the poles of the function $F(s)$ means the zeros of $D(s)$ are simple roots. This means the roots of $D(s)$ are all of order one. In other words, they can only return zero value for the $D(s)$ itself and cannot make the derivations of $D(s)$ zero. Some of these roots may be real and the others may be complex. The real and complex roots are shown with r_i and r_c, respectively. Also, the number of real and complex roots are n_r and n_c, respectively. In the case of possessing only simple poles, the inverse Laplace transformation of the function $F(s)$ in the frequency domain can be calculated via:

$$f(t) = \mathcal{L}^{-1}\{F(s)\} = \text{Re}\left(\sum_{i=1}^{n_c} \frac{N(c_i)}{E(c_i)} e^{c_i t} \right) + \sum_{j=1}^{n_r} \frac{N(r_j)}{E(r_j)} e^{r_j t} \qquad (2.233)$$

in which the function $E(s)$ is the first derivative of the function $D(s)$ with respect to variable s. It is notable that the sign $\text{Re}(.)$ returns the real part of any complex input which is inserted in the parenthesis. Obviously, once all increments of the aforementioned procedure are completed, the time-dependent part of the displacement vector of the plate under such a loading is obtained. Thereafter, the time-dependent parts will be substituted in the Eq. (2.221) to reach an analytical solution for both bending and shear deflections of the plate.

2.3   HOMOGENIZATION OF NANOCOMPOSITES

In the past two sections, the tools for reaching the governing equations of nanocomposite structures were presented to enable the readers to derive the governing equations of beams, plates, and shells under defined working conditions. However, in Section 2.2 some integrals were defined either over the thickness or over the cross-sectional area of the observed structure. Determining these integrals is impossible unless the material properties of the constituent material is in hand. Therefore, in this section the issue of reaching the equivalent material properties of various types of nanocomposites employed in this book will be depicted to give insight about how to develop the integrals previously defined in Section 2.2. Since this book investigates the mechanical responses of different types of nanocomposite materials such as CNTR, GPLR, and GOR materials, the homogenization procedure for all of these types of nanocomposites will be discussed in this section in detail. In addition, the effect of single- or multi-layered structures will be covered. For the sake of simplicity, homogenization will be explained for single-layered nanocomposite materials and afterward it will be shown how to derive the material properties for multi-layered ones.

2.3.1 MATERIAL PROPERTIES OF SINGLE-LAYERED CNTR NANOCOMPOSITES

In this section the concept of the well-known rule of the mixture will be implemented for the goal of reaching the equivalent material properties of nanocomposites reinforced with CNTs. In this approach, the effect of the small size of the CNTs on the effective material properties is covered using a group of scale-dependent coefficients. Such coefficients were derived by comparing the experimental data with those achieved from MD simulations. According to this approach, the equivalent stiffness of the CNTR nanocomposite can be calculated using the following formula:

$$E_{11} = \eta_1 V_{CNT} E_{11}^{CNT} + V_m E^m,$$

$$\frac{\eta_2}{E_{22}} = \frac{V_{CNT}}{E_{22}^{CNT}} + \frac{V_m}{E^m},$$

$$\frac{\eta_3}{G_{12}} = \frac{V_{CNT}}{G_{12}^{CNT}} + \frac{V_m}{G^m}$$

$$(2.234)$$

where E_{11}, E_{22}, and G_{12} are longitudinal, transverse, and shear modules of the CNTR nanocomposite, respectively. Also, E_{11}^{CNT}, E_{22}^{CNT}, and G_{12}^{CNT} are the longitudinal, transverse, and shear modules of the CNTs, respectively. The matrix is assumed to be an isotropic polymeric material with elastic moduli of E^m and shear moduli of G^m. In the equation above, the size-dependent coefficients provided to capture the small size of the CNTs are η_i, $i = (1,2,3)$. These coefficients vary as the volume fraction of the CNTs changes. The equation above can predict the shear moduli in the xy plane and the value of this moduli in xz and yz planes can be assumed to be 0.5 times of the value achieved for the shear moduli in the xy plane ($G_{13} = G_{23} = 0.5G_{12}$). The volume fraction of the CNTs in the matrix is shown with V_{CNT} in the previous relations and it can be related to the volume fraction of the matrix (V_m) via:

$$V_{CNT} + V_m = 1$$

$$(2.235)$$

The volume fraction of the CNTs in the nanocomposite can be computed using the following formula:

$$V_{CNT} = \frac{W_{CNT}}{W_{CNT}\left(1 - \dfrac{\rho_{CNT}}{\rho_m}\right) + \dfrac{\rho_{CNT}}{\rho_m}}$$

$$(2.236)$$

In the equation above, W_{CNT} is the weight fraction of the CNTs in the nanocomposite. It is noteworthy that the mass density of the matrix and CNT are shown with ρ_m

and ρ_{CNT}, respectively. The previous formula can provide the uniformly distributed (UD) CNTs across the thickness of the structure; however, the FG distribution of the CNTs can be attained by defining a function to handle the distribution of the volume fraction of the nanofillers over the thickness of the structure. To obtain FG nanocomposites, the following changes must be made in the volume fraction of the CNTs inside the nanocomposite:

$$V_{CNT} = \begin{cases} \dfrac{|4z|}{h}\dfrac{W_{CNT}}{W_{CNT}\left(1-\dfrac{\rho_{CNT}}{\rho_m}\right)+\dfrac{\rho_{CNT}}{\rho_m}} & , & \text{FG-X} \\[4ex] \left[2-\dfrac{|4z|}{h}\right]\dfrac{W_{CNT}}{W_{CNT}\left(1-\dfrac{\rho_{CNT}}{\rho_m}\right)+\dfrac{\rho_{CNT}}{\rho_m}} & , & \text{FG-O} \\[4ex] \left[2+\dfrac{4z}{h}\right]\dfrac{W_{CNT}}{W_{CNT}\left(1-\dfrac{\rho_{CNT}}{\rho_m}\right)+\dfrac{\rho_{CNT}}{\rho_m}} & , & \text{FG-V} \end{cases} \tag{2.237}$$

Furthermore, the equivalent Poisson's ratio and mass density of the CNTR nanocomposites can be calculated using the following formula:

$$v_{12} = V_{CNT}v_{12}^{CNT} + V_m v_m, \tag{2.238}$$

$$\rho = V_{CNT}\rho_{CNT} + V_m\rho_m \tag{2.239}$$

Now all of the material properties required to reach the inertia and rigidity integrals are in hand. As can be realized, the material properties of the CNTR nanocomposite belong to an orthotropic media. Therefore, the constitutive equation for such a nanocomposite can be presented in the following form:

$$\begin{bmatrix} \sigma_{xx} \\ \sigma_{yy} \\ \sigma_{xy} \\ \sigma_{xz} \\ \sigma_{yz} \end{bmatrix} = \begin{bmatrix} Q_{11} & Q_{12} & 0 & 0 & 0 \\ Q_{21} & Q_{22} & 0 & 0 & 0 \\ 0 & 0 & Q_{66} & 0 & 0 \\ 0 & 0 & 0 & Q_{55} & 0 \\ 0 & 0 & 0 & 0 & Q_{44} \end{bmatrix} \begin{bmatrix} \varepsilon_{xx} \\ \varepsilon_{yy} \\ \gamma_{xy} \\ \gamma_{xz} \\ \gamma_{yz} \end{bmatrix} \tag{2.240}$$

where

$$Q_{11} = \frac{E_{11}}{1-v_{12}v_{21}}, \quad Q_{12} = \frac{v_{12}E_{22}}{1-v_{12}v_{21}}, \quad Q_{22} = \frac{E_{22}}{1-v_{12}v_{21}}, \tag{2.241}$$

$$Q_{44} = G_{23}, \quad Q_{55} = G_{13}, \quad Q_{66} = G_{12}$$

2.3.2 Material Properties of Single-Layered GPLR Nanocomposites

In this section the development of the effective material properties of GPLR nanocomposite materials will be explained. To reach the material properties of such polymeric nanocomposites, the Halpin-Tsai micromechanical method will be employed to cover the influences of the nanofillers' small dimensions and particular shape on the material properties of the nanocomposite under observation. According to this method, the equivalent stiffness of the GPLR nanocomposite can be determined as follows:

$$E_e = \frac{3}{8} E_\Lambda + \frac{5}{8} E_\Theta \tag{2.242}$$

where E_Λ and E_Θ are longitudinal and transverse modules in x and y directions, respectively. These modules can be defined as follows:

$$E_\Lambda = \frac{1 + \xi_\Lambda \eta_\Lambda V_{GPL}}{1 - \eta_\Lambda V_{GPL}} E_m,$$

$$E_\Theta = \frac{1 + \xi_\Theta \eta_\Theta V_{GPL}}{1 - \eta_\Theta V_{GPL}} E_m \tag{2.243}$$

where

$$\eta_\Lambda = \frac{(E_{GPL} / E_m) - 1}{(E_{GPL} / E_m) + \xi_\Lambda},$$

$$\eta_\Theta = \frac{(E_{GPL} / E_m) - 1}{(E_{GPL} / E_m) + \xi_\Theta} \tag{2.244}$$

In the equations above, E_{GPL} and E_m denote the Young's moduli of GPLs and polymeric matrix, respectively. Also, the volume fraction of the GPLs in the nanocomposite is shown with V_{GPL}. The Halpin-Tsai method includes the impact of the particular geometry of the reinforcements on the mechanical properties of the nanocomposite material. These geometrical parameters can be determined as follows:

$$\xi_\Lambda = \frac{2 l_{GPL}}{h_{GPL}},$$

$$\xi_\Theta = \frac{2 w_{GPL}}{h_{GPL}} \tag{2.245}$$

In the equation above, l_{GPL}, w_{GPL}, and h_{GPL} are the average length, width, and thickness of the GPLs, respectively. Moreover, the equivalent The terms which are provided to account for the geometrical shape of the GPLs are ξ_Λ and ξ_Θ. Poisson's ratio, mass density, and CTE of the GPLR nanocomposite material can be estimated via the simple form of the rule of the mixture as follows:

$$\rho_e = \rho_{GPL} V_{GPL} + \rho_m V_m, \tag{2.246}$$

$$\nu_e = \nu_{GPL} V_{GPL} + \nu_m V_m, \tag{2.247}$$

$$\alpha_e = \alpha_{GPL} V_{GPL} + \alpha_m V_m \tag{2.248}$$

In the equations above, the volume fraction of the GPLs can be presented in the following form:

$$V_{GPL} = \frac{W_{GPL}}{W_{GPL}\left(1 - \dfrac{\rho_{GPL}}{\rho_m}\right) + \dfrac{\rho_{GPL}}{\rho_m}} \tag{2.249}$$

where W_{GPL} is the weight fraction of the GPLs. It must be stated that the previous formula reveals the volume fraction of the GPLs for the UD pattern. Once the FG pattern is desired, the following formulas must be implemented for the volume fraction of the GPLs:

$$V_{GPL} = \begin{cases} \dfrac{|4z|}{h} \dfrac{W_{GPL}}{W_{GPL}\left(1 - \dfrac{\rho_{GPL}}{\rho_m}\right) + \dfrac{\rho_{GPL}}{\rho_m}} & , \quad \text{FG-X} \\[20pt] \left[2 - \dfrac{|4z|}{h}\right] \dfrac{W_{GPL}}{W_{GPL}\left(1 - \dfrac{\rho_{GPL}}{\rho_m}\right) + \dfrac{\rho_{GPL}}{\rho_m}} & , \quad \text{FG-O} \\[20pt] \left[2 + \dfrac{4z}{h}\right] \dfrac{W_{GPL}}{W_{GPL}\left(1 - \dfrac{\rho_{GPL}}{\rho_m}\right) + \dfrac{\rho_{GPL}}{\rho_m}} & , \quad \text{FG-V} \end{cases} \tag{2.250}$$

Now the constitutive equations of GPLR nanocomposite materials can be written to show the Q_{ij} coefficients employed in Section 2.2 for calculating the rigidity integrals. As can be seen from the derived material properties, the implemented homogenization procedure achieves the material properties of an isotropic nanocomposite material. The constitutive equations of GPLR nanocomposites are entirely as same as that introduced in Eq. (2.240). However, the components of the stiffness matrix of GPLR nanocomposites are in the following form:

$$Q_{11} = \frac{E_e}{1 - \nu_e^2}, \quad Q_{12} = \nu_e Q_{11}, \quad Q_{22} = Q_{11},$$
$$Q_{44} = G_e, \quad Q_{55} = G_e, \quad Q_{66} = G_e \tag{2.251}$$

In Eq. (2.251), G_e stands for the shear moduli of the GPLR nanocomposite which can be obtained using the definition of the shear moduli for linear elastic isotropic materials.

2.3.3 MATERIAL PROPERTIES OF SINGLE-LAYERED GOR NANOCOMPOSITES

In this section, we will discuss extracting the equivalent material properties of GOR nanocomposite materials. The derivation procedure will be founded on the basis of the well-known Halpin-Tsai micromechanical scheme. Again, remember that the impact of the geometry of the implemented nanoparticles will be included, too. According to the Halpin-Tsai method, the stiffness of the GOR nanocomposite can be presented in the following form:

$$E_e = 0.49E_\Lambda + 0.51E_\Theta \tag{2.252}$$

where E_Λ and E_Θ are longitudinal and transverse modules in x and y directions, respectively. Based on the Halpin-Tsai method, the longitudinal and transverse modules can be defined as follows:

$$E_\Lambda = \frac{1+\xi_\Lambda \eta_\Lambda V_{GO}}{1-\eta_\Lambda V_{GO}} E_m,$$

$$E_\Theta = \frac{1+\xi_\Theta \eta_\Theta V_{GO}}{1-\eta_\Theta V_{GO}} E_m \tag{2.253}$$

in which

$$\eta_\Lambda = \frac{(E_{GO} / E_m)-1}{(E_{GO} / E_m)+\xi_\Lambda},$$

$$\eta_\Theta = \frac{(E_{GO} / E_m)-1}{(E_{GO} / E_m)+\xi_\Theta} \tag{2.254}$$

and in which E_{GO} and E_m denote the Young's moduli of the GO and matrix, respectively. Also, the geometrical coefficients (ξ_Λ and ξ_Θ) will be in the following form for the GO reinforcement:

$$\xi_\Lambda = \xi_\Theta = \frac{2 d_{GO}}{h_{GO}} \tag{2.255}$$

where d_{GO} and h_{GO} are the average diameter and thickness of the GO, respectively. Now, the equivalent Poisson's ratio and mass density of the GOR nanocomposite can be obtained via the following formula based on the rule of the mixture:

$$v_e = v_{GO}V_{GO} + v_m V_m, \tag{2.256}$$

$$\rho_e = \rho_{GO}V_{GO} + \rho_m V_m \tag{2.257}$$

$$\rho_e = \rho_{GO}V_{GO} + \rho_m V_m \tag{2.258}$$

Also, the equivalent CTE of the GOR nanocomposite can be developed using the following formula:

$$\alpha_e = \alpha_m + \frac{\alpha_m + \alpha_{GO}}{\dfrac{1}{K_m} + \dfrac{1}{K_{GO}}} \left[\frac{1}{K_e} + \frac{1}{K_m} \right] \qquad (2.259)$$

where α_m and α_{GO} are the CTE of the matrix and GOs, respectively. Moreover, the bulk moduli of matrix and GOs were shown with K_m and K_{GO}, respectively. Clearly, the equivalent bulk moduli of the GOR nanocomposite was shown with K_e. In all of the equations above, the volume fraction of the GOs can be achieved using the following relation:

$$V_{GO} = \frac{W_{GO}}{W_{GO}\left(1 - \dfrac{\rho_{GO}}{\rho_m}\right) + \dfrac{\rho_{GO}}{\rho_m}} \qquad (2.260)$$

Again, it must be stated that the equation above states the volume fraction of the GOs once UD of GOs across the thickness of the structure is utilized. On the other hand, once the GOs are scattered across the thickness of the structure via a desired shape (i.e., FG nanocomposites), the volume fraction of the GOs must be formulated in the following form:

$$V_{GO} = \begin{cases} \dfrac{|4z|}{h} \dfrac{W_{GO}}{W_{GO}\left(1 - \dfrac{\rho_{GO}}{\rho_m}\right) + \dfrac{\rho_{GO}}{\rho_m}} & , \quad \text{FG-X} \\[4ex] \left[2 - \dfrac{|4z|}{h}\right] \dfrac{W_{GO}}{W_{GO}\left(1 - \dfrac{\rho_{GO}}{\rho_m}\right) + \dfrac{\rho_{GO}}{\rho_m}} & , \quad \text{FG-O} \\[4ex] \left[2 + \dfrac{4z}{h}\right] \dfrac{W_{GO}}{W_{GO}\left(1 - \dfrac{\rho_{GO}}{\rho_m}\right) + \dfrac{\rho_{GO}}{\rho_m}} & , \quad \text{FG-V} \end{cases} \qquad (2.261)$$

Now one can calculate the components of the elasticity tensor for a GOR nano-composite material to make the rigidities. Indeed, the constitutive equations of such nanocomposite materials can be made similarly to that of a linear elastic isotropic material. The stress-strain relationship for GOR nanocomposites can be considered to be in the form introduced in Eq. (240). Also, the components of the elasticity tensor (i.e., Q_{ij}'s) can be calculated the same as Eq. (2.251). However, instead of the equivalent modules and Poisson's ratio of the GPLR nanocomposites, those of GOR nanocomposites must be replaced.

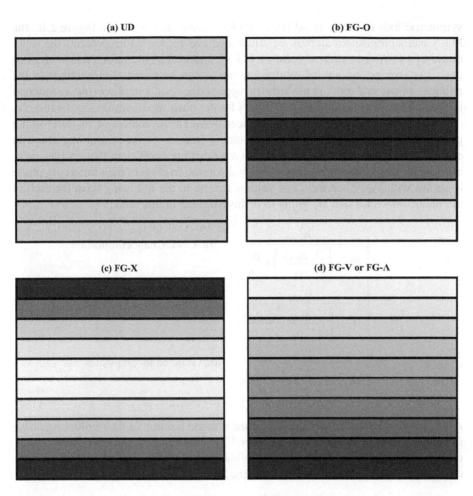

FIGURE 2.1 Schematic of various distribution patterns of nanofillers or nanoparticles across the thickness direction for a ten-layered nanocomposite material. The distributions shown in this section can be applied for any desired multi-layered nanocomposite.

2.3.4 Material Properties of Multi-Layered Nanocomposites

In the previous parts (i.e., Sections 2.3.1, 2.3.2, and 2.3.3), the derivation of the equivalent material properties of CNTR, GPLR, and GOR nanocomposite materials was explained in detail for the case of analyzing a single-layered nanocomposite. However, in some applications, multi-layered (sandwich) nanocomposite materials will be utilized in the mechanical analysis of nanocomposite structures. This section will show the effects of more than one layer of nanocomposite on the effective material properties of the multi-layered nanocomposite materials.

Consider an N-layered nanocomposite material, shown in Figure 2.1. Figure 2.1a illustrates the case of having identical content of reinforcing phase in each of the layers (i.e., UD nanocomposites). Figure 2.1b reveals nanocomposites that possess the maximum amount of reinforcing phase in the surfaces close to the horizontal symmetric axis of the material (i.e., the FG-O nanocomposites). In Figure 2.1c, the FG-X nanocomposite materials are illustrated and Figure 2.1d represents the schematic of FG-Λ (i.e., reverse form of FG-V) nanocomposite materials. In the last case, the greatest amount of reinforcing phase can be found at the bottom of the nanocomposite and moving towards the upper edge results in observing a reduction in the content of the employed nanofillers for the purpose of reinforcing the matrix.

To include the impact of the differences between the content of the reinforcing phases used in each of the layers, the volume fraction of the reinforcing phase in each of the layers must be multiplied by a modifying coefficient that enables us to account for the difference. Assume the total volume fraction of the reinforcing phase is shown with V_{RP}. Therefore, this volume can be in the following form for each of the nanocomposites that are going to be investigated in this book:

$$
V_{RP} = \begin{cases}
\dfrac{W_{CNT}}{W_{CNT}\left(1 - \dfrac{\rho_{CNT}}{\rho_m}\right) + \dfrac{\rho_{CNT}}{\rho_m}} & \text{for CNTR nanocomposites} \\[3em]
\dfrac{W_{GPL}}{W_{GPL}\left(1 - \dfrac{\rho_{GPL}}{\rho_m}\right) + \dfrac{\rho_{GPL}}{\rho_m}} & \text{for GPLR nanocomposites} \\[3em]
\dfrac{W_{GO}}{W_{GO}\left(1 - \dfrac{\rho_{GO}}{\rho_m}\right) + \dfrac{\rho_{GO}}{\rho_m}} & \text{for GOR nanocomposites}
\end{cases}
\tag{2.262}
$$

Also, consider the minimum content of the volume fraction of the reinforcing phase to be $V_{\min}$ and its maximum content to be $V_{\max}$. Once each of the distributions illustrated in Figure 2.1 are employed, the following expressions must be considered for the volume fraction available in each of the layers of the material (V_i):

$$
V_i = \frac{V_{RP}}{N} \qquad \text{for UD,}
\tag{2.263}
$$

$$
V_i = \begin{cases}
V_{\min} + (i-1)\dfrac{2(V_{\max} - V_{\min})}{N-2}, & i \le N/2 \\[2em]
V_{\max} - \left(i - \dfrac{N}{2} - 1\right)\dfrac{2(V_{\max} - V_{\min})}{N-2}, & i > N/2
\end{cases}
\qquad \text{for FG-O,}
\tag{2.264}
$$

$$
V_i = \begin{cases}
V_{\max} - (i-1)\dfrac{2(V_{\max} - V_{\min})}{N-2}, & i \le N/2 \\[2em]
V_{\min} + \left(i - \dfrac{N}{2} - 1\right)\dfrac{2(V_{\max} - V_{\min})}{N-2}, & i > N/2
\end{cases}
\qquad \text{for FG-X,}
\tag{2.265}
$$

$$
V_i = V_{\min} + (i-1)\frac{V_{\max} - V_{\min}}{N-1} \qquad \text{for FG-Λ or FG-V}
\tag{2.266}
$$

Consider that in the case of analyzing a multi-layered nanocomposite structure, all of the mass moments of inertia and cross-sectional rigidities defined in Section 2.2 must be calculated using summation over the layers. This issue will be explained in the framework of a pure mathematical example. Assume the following integral to be valid across the thickness:

$$\Xi = \int_{-\frac{h}{2}}^{\frac{h}{2}} F(z)dz \tag{2.267}$$

In the equation above, the function $F(z)$ is continuous over the range $\left(-\frac{h}{2}, \frac{h}{2}\right)$. Now, assume that the magnitude of this function varies across the thickness and its magnitude depends on the subdomains in the following form:

$$F(z) = \begin{cases} F_1 & , & -\dfrac{h}{2} < z < -\dfrac{h}{2} + \Delta h \\[2mm] F_2 & , & -\dfrac{h}{2} + \Delta h < z < -\dfrac{h}{2} + 2\Delta h \\[2mm] \vdots & , & \vdots \\[2mm] F_{N-1} & , & \dfrac{h}{2} - 2\Delta h < z < \dfrac{h}{2} - \Delta h \\[2mm] F_N & , & \dfrac{h}{2} - \Delta h < z < \dfrac{h}{2} \end{cases} \tag{2.268}$$

In this case the integral presented in Eq. (2.267) must be calculated in the following form:

$$\Xi = \sum_{i=1}^{N} \int_{\frac{h}{2} - i\Delta h}^{\frac{h}{2} - (i-1)\Delta h} F_i(z)dz \tag{2.269}$$

So, the cross-sectional rigidities and the mass moments of inertia must be calculated for an N-layered FG nanocomposite using the framework explained in the equations above.

2.3.5 MATERIAL PROPERTIES OF POROUS SINGLE-LAYERED NANOCOMPOSITES

In this section the influence of the existence of porosities in the nanocomposite material on the mechanical properties of the aforementioned materials will be discussed. In other words, a mathematical representation of the issue of availability of porosities in nanocomposites will be presented. In what follows, the effects of three types of porosities on the material properties will be covered. In porosity type-I, the size of the existing pores will become larger when moving toward the mid-plane of the structure from its top or bottom edges. The reverse trend can be observed in porosity type-II. In other words, the size of pores in the top and bottom edges are bigger than those in the mid-plane. Finally, uniform size pores can be seen in porosity type-III. The schematic of these three types of porous materials can be observed in Wang et al. (2019). The equivalent material properties of porous nanocomposite materials can be presented in the following form (Wang et al., 2019):

$$
E(z) = \begin{cases} E^*\left[1 - e_1 \cos\left(\dfrac{\pi z}{h}\right)\right] & \text{Porosity type-I} \\[3mm] E^*\left[1 - e_2\left\{1 - \cos\left(\dfrac{\pi z}{h}\right)\right\}\right] & \text{Porosity type-II} \\[3mm] \rho^* e_3 & \text{Porosity type-III} \end{cases}
\tag{2.270}
$$

$$
\rho(z) = \begin{cases} \rho^*\left[1 - e_{m1} \cos\left(\dfrac{\pi z}{h}\right)\right] & \text{Porosity type-I} \\[3mm] \rho^*\left[1 - e_{m2}\left\{1 - \cos\left(\dfrac{\pi z}{h}\right)\right\}\right] & \text{Porosity type-II} \\[3mm] \rho^* e_{m3} & \text{Porosity type-III} \end{cases}
\tag{2.271}
$$

$$
v(z) = v^*
\tag{2.272}
$$

in which E^*, ρ^*, and v^* are the Young's moduli, mass density, and Poisson's ratio of the non-porous isotropic nanocomposite, respectively. This type of porosity-based method can be utilized for both GOR and GPLR nanocomposite materials that were homogenized in the previous sections with the assumption of being isotropic linear elastic solids. The relation between the equivalent Young's moduli and mass density is in the following form (Wang et al., 2019):

$$
\frac{E(z)}{E^*} = \left[\frac{\rho(z)}{\rho^*}\right]^2
\tag{2.273}
$$

Following Eq. (2.273), the relations between the porosity coefficients (e_1, e_2, e_3) and mass density coefficients (e_{m1}, e_{m2}, e_{m3}) can be formulated in the following form:

$$
\begin{cases} 1 - e_{m1} \cos\left(\dfrac{\pi z}{h}\right) = \sqrt{1 - e_1 \cos\left(\dfrac{\pi z}{h}\right)} \\[3mm] 1 - e_{m2}\left[1 - \cos\left(\dfrac{\pi z}{h}\right)\right] = \sqrt{1 - e_2\left[1 - \cos\left(\dfrac{\pi z}{h}\right)\right]} \\[3mm] e_{m3} = \sqrt{e_3} \end{cases}
\tag{2.274}
$$

Furthermore, during these mathematical manipulations, the consistency of the equivalent mass of the porous nanocomposite is one of the assumptions which must be satisfied. To this purpose, the following relation can be written involving all three types of porous nanocomposite materials:

$$\int_0^{h/2} \sqrt{1 - e_1 \cos\left(\frac{\pi z}{h}\right)}\, dz = \int_0^{h/2} \sqrt{1 - e_2\left[1 - \cos\left(\frac{\pi z}{h}\right)\right]}\, dz = \int_0^{h/2} \sqrt{e_3}\, dz \qquad (2.275)$$

According to the equation above, the porosity coefficients e_2 and e_3 can be found once the coefficient e_1 is known.

In addition to the effects of existence of porosities in the media, the effect of various distributions for the reinforcing elements across the thickness of the nano-composite is covered, too. Indeed, three types of nanocomposites will be discussed, namely type-A, type-B, and type-C. In type-A nanocomposites, the maximum amount of the nanosize reinforcements can be found in the top and bottom edges and no reinforcing element can be found in the mid-plane; whereas, the distribution of the reinforcements is completely opposite in type-B nanocomposites. Type-C nano-composites contain a constant content of the reinforcements in all spatial positions across the thickness of the material. Due to this variety, the volume fraction of the reinforcements possesses a thickness-dependent formula for nanocomposites of type A and B. Hence, the volume fraction of the reinforcing phase can be presented in the following form for each type of nanocomposite:

$$V_{NSR}(z) = \begin{cases} s_{i1}\left[1 - \cos\left(\frac{\pi z}{h}\right)\right] & \text{Type-A} \\[2em] s_{i2} \cos\left(\frac{\pi z}{h}\right) & \text{Type-B} \\[2em] s_{i3} & \text{Type-C} \end{cases} \qquad (2.276)$$

In the equation above, V_{NSR} stands for the volume fraction of the nanosize reinforce-ment (NSR). The coefficients s_{ij}, $j = 1, 2, 3$ can be determined using the following definition:

$$V_T \int_{-h/2}^{h/2} \frac{\rho(z)}{\rho^*}\, dz = \begin{cases} s_{i1} \int_{-h/2}^{h/2}\left[1 - \cos\left(\frac{\pi z}{h}\right)\right]\frac{\rho(z)}{\rho^*}\, dz \\[2em] s_{i2} \int_{-h/2}^{h/2} \cos\left(\frac{\pi z}{h}\right)\frac{\rho(z)}{\rho^*}\, dz \\[2em] s_{i3} \int_{-h/2}^{h/2} \frac{\rho(z)}{\rho^*}\, dz \end{cases} \qquad (2.277)$$

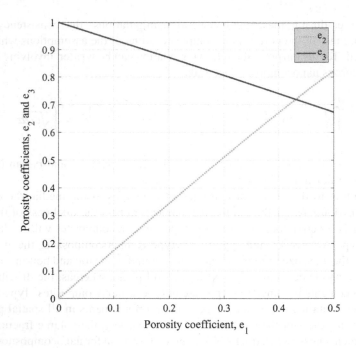

FIGURE 2.2 Variation of the second and third porosity coefficients against first porosity coefficient for a GOR polymeric nanocomposite material.

In the equation above, V_T stands for the total volume fraction of the reinforcement. This term can be derived using Eqs. (2.249) and (2.260) for GPLR and GOR nanocomposites, respectively. The derivation of the equivalent material properties is now finished. It might be a little confusing to understand how the porosity coefficients can be related to each other. Therefore, it is decided to present the variation of the porosity coefficients, e_2 and e_3, in terms of e_1 for the sake of clarity. This issue is studied in the framework of Figure 2.2. It can be realized that the porosity coefficient, e_2, behaves similar with e_1 and it will be enlarged once a greater value is assigned to e_1. However, the third porosity coefficient will be lessened as the coefficient e_1 is added.

Also, the influence of the porosity coefficient e_1 on the variation of the mass coefficients for any desired dimensionless thickness is illustrated in Figure 2.3. Based on this figure, the value of the mass coefficients e_{m1} and e_{m3} will be increased as the porosity coefficient e_1 is added; whereas, the mass coefficient e_{m2} will be lessened once the porosity coefficient is amplified. Moreover, it is clear that the maximum amount of the mass coefficient for a type-I distribution is in the mid-plane of the structure, while, this maximum amount can be found at the top and bottom edges once the type-II distribution of the porosities is investigated.

2.3.6 Material Properties of Viscoelastic Nanocomposites

In this section the influence of the viscoelastic behavior of the polymeric nanocomposite materials will be investigated to reach the constitutive equations of

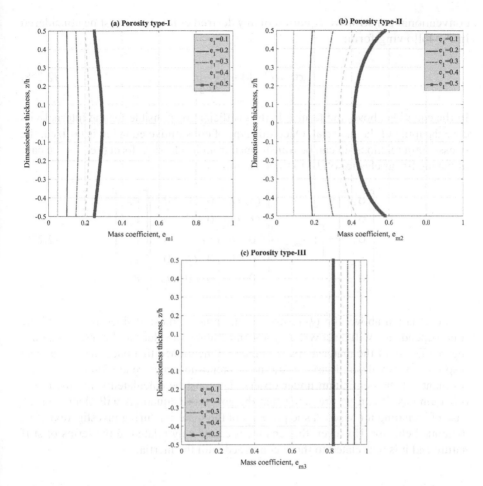

FIGURE 2.3 Effect of the dimensionless thickness on the mass coefficients of a GOR polymeric nanocomposite material once the porosity coefficient is varied.

nanocomposites once viscoelasticity effects are included. Viscoelastic materials possess material properties depending on the time variable. Hence, the constitutive equations of linear elastic solids cannot be employed when analyzing a viscoelastic material. Many models can be found in the literature that enable us to consider the time-dependency of the material properties (Brinson and Brinson, 2015). It is ideal to determine the viscoelastic properties of any desired material using dynamic mechanical analysis (DMA) and experimental approaches. However, doing experimental creep and relaxation tests may not be possible in any given condition. Therefore scientists made some mathematical efforts to present a rough underestimation of the material's behavior. In this section, the well-known three-parameter model will be implemented to cover the time-dependency of the material properties. According to this model, the conventional constitutive equation of any desired continuum must be considered in the following form:

$$\sigma_{ij} = C_{ijkl}\left[1 + g\frac{\partial}{\partial t}\right]\varepsilon_{kl} \tag{2.278}$$

In the equation above, the term g is the coefficient responsible for the internal viscose damping of the material. Once this type of constitutive equation is utilized, the stress-strain relationship can be transformed from its classical form (i.e., for the case of having linear elastic solids) to the following one:

$$\begin{bmatrix} \sigma_{xx} \\ \sigma_{yy} \\ \sigma_{yz} \\ \sigma_{xz} \\ \sigma_{xy} \end{bmatrix} = \left[1 + g\frac{\partial}{\partial t}\right]\begin{bmatrix} Q_{11} & Q_{12} & 0 & 0 & 0 \\ Q_{12} & Q_{22} & 0 & 0 & 0 \\ 0 & 0 & Q_{44} & 0 & 0 \\ 0 & 0 & 0 & Q_{55} & 0 \\ 0 & 0 & 0 & 0 & Q_{66} \end{bmatrix}\begin{bmatrix} \varepsilon_{xx} \\ \varepsilon_{yy} \\ \varepsilon_{yz} \\ \varepsilon_{xz} \\ \varepsilon_{xy} \end{bmatrix} \tag{2.279}$$

In the equation above, the Q_{ij} terms are the same as those defined before and the time-dependency will be covered by assigning nonzero values to the internal damping coefficient of the material. Now, it must be mentioned that once the mechanical responses of any desired viscoelastic nanocomposite structure are found, the stress resultants of the continuum under observation must be calculated using the previous formulas. It can be predicted that the governing equations will change in the case of including the viscoelastic property of the material during investigation of its dynamic behavior. However, this change is completely related to the terms of stiff nature and it is not related to those concerned with the inertia.

Part 2

Static Analyses

3 Buckling Analysis of Embedded GOR Nanocomposite Shells

3.1 PROBLEM DEFINITION

In this chapter, the static buckling responses of uniformly distributed graphene oxide reinforced (UD-GOR) nanocomposite cylindrical shells will be examined to describe the reaction of this type of nanocomposite structures when subjected to a compressive buckling load. Figure 3.1 shows the geometry and coordinate system used to solve the problem.

The structure is assumed to rest on an elastic medium with two springs, namely Winkler and Pasternak springs. The problem is formulated using the FSDT of cylindrical shells to consider the effects of shear deformation on the buckling behaviors of the shell up to first-order. The governing equations of the problem are shown in Eqs. (2.77)–(2.81). However, in the referenced set of governing equations, the influences of time-dependent terms must be ignored to reach the governing equations for the static buckling problem. It must also be noted that there are no radial and lateral forces applied on the shell and these terms must be set to zero as well. Solving the equations reveals the critical buckling load of the nanocomposite shell. In what follows, a group of numerical illustrations will be presented for the purpose of discussing variants that can affect the buckling behaviors of the GOR nanocomposite structures.

3.2 NUMERICAL RESULTS AND DISCUSSION

The following numerical examples will illustrate the qualitative and quantitative influences of various parameters on the buckling response of the GOR nanocomposite shells. The nanocomposite material is assumed to be made from an epoxy matrix that hosts the reinforcing GOs. The material properties of the matrix and nanosize reinforcements are from Zhang et al. (2018). It is best to introduce the dimensionless form of the buckling load and foundation parameters utilized in the following results before starting to discuss these issues:

$$N_{cr} = \frac{N^b}{100E_m h^2 / R\sqrt{3(1-v_m)}}, K_w = \frac{k_w L^4}{E_m h^3 / 12(1-v_m^2)}, K_p = \frac{k_p L^2}{E_m h^3 / 12(1-v_m^2)} \quad (3.1)$$

where L and R denote the length and radius of the cylindrical shell, respectively. Also, E_m and v_m account for the Young's moduli and Poisson's ratio of the matrix, respectively.

Figure 3.2 shows the variation of the first dimensionless buckling load of the GOR nanocomposite cylindrical shell versus circumferential wave number once

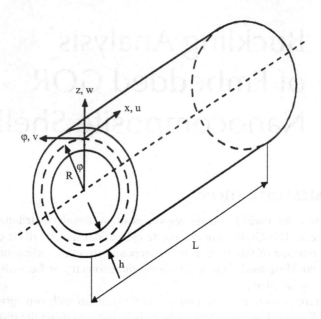

FIGURE 3.1 The geometry and coordinate system of a cylindrical shell.

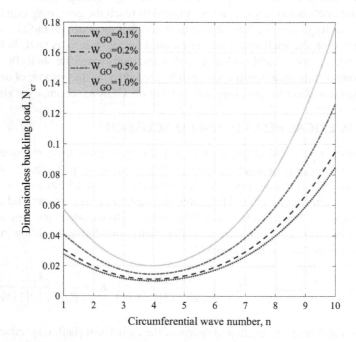

FIGURE 3.2 Variation of the dimensionless buckling load of S-S nanocomposite shells against circumferential wave number for various weight fractions of GOs ($L/h = R/h = 20$, $K_w = 100$, $K_p = 10$).

different contents of reinforcing GOs are implemented. Observe that the buckling load of the shell will shift upward as the content of the GO grows in the composition of the nanocomposite. The reason for this phenomenon is that the equivalent stiffness of the nanocomposite material improves when the weight fraction of the GOs increases, which leads to a corresponding increase in the cross-sectional rigidities of the shell. It is obvious that enhancement of the cross-sectional rigidities of the shell results in a higher stiffness matrix for the structure, and, due to the direct relation between stiffness and buckling load, the cylinder will be able to tolerate greater buckling loads.

The influences of foundation parameters on the critical buckling load of the GOR nanocomposite shells are covered in Figures 3.3a and 3.3b for two cases using 0.1% and 0.5% weight fractions for the GOs. It is clear that the mechanical endurance of the shell grows as the stiffness of the elastic springs employed in the medium is intensified. As explained in the interpretation of the former illustration, the buckling load of the shell possesses a direct relation with the equivalent stiffness of the structure. Therefore, it is not strange to see an increase in the buckling load of the shell once the stiffness of the foundation intensifies. It is noteworthy that the impact of the Pasternak spring is more powerful than that of the Winkler's. This reality could be estimated by taking a brief look at the governing equations of the shell. The Pasternak springs are stiffer than Winkler springs, so, if an identical dimensionless value is assigned to the dimensionless coefficients of these parameters, the Pasternak coefficient can induce more changes in the mechanical response of the structure compared with the Winkler coefficient.

In Figure 3.4, the coupled impacts of changing the weight fraction of the GOs and using various radius-to-thickness ratios for the shell on the dimensionless critical buckling response of the cylinder are covered. It is clear that the GOR nanocomposite shells with larger amounts of GOs can endure greater static forces without reaching the buckled state. It was previously mentioned that increasing the weight fraction of the GOs results in a higher equivalent stiffness for the structure, which leads to an enhancement in the critical buckling load of the shell-type element. On the other hand, except for a limited range of circumferential wave numbers (i.e., approximately from 1 to 4), the buckling load of the structure will go through a reducing path as the radius-to-thickness ratio becomes greater. The reason for this is that in greater radius-to-thickness ratios the structure moves toward being thinner and this can intensify the flexibility of the shell. As we know, any increase in the flexibility of the structure corresponds with a reciprocal reduction in the stiffness of the same media. Therefore, the nanocomposite structures can endure lower buckling loads once they are thinner.

Figure 3.5 displays the combined impacts of length-to-thickness ratio of the shell and the weight fraction of the GOs on the buckling behaviors of GOR nanocomposite shells. Looking at this diagram, it can be seen that the buckling load of the cylindrical shell can be greatly increased as the length-to-thickness ratio of the structure grows.

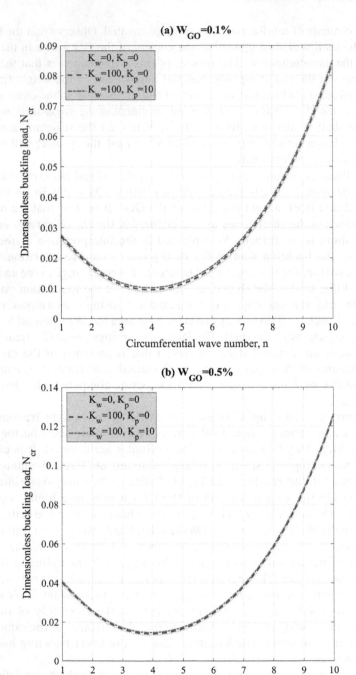

FIGURE 3.3 Variation of the dimensionless buckling load of S-S nanocomposite shells versus circumferential wave number for various foundation parameters at (a) $W_{GO} = 0.1\%$ and (b) $W_{GO} = 0.5\%$ ($L/h = R/h = 20$).

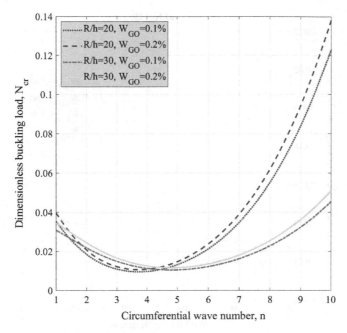

FIGURE 3.4 Variation of the dimensionless buckling load of S-S nanocomposite shells versus circumferential wave number for different radius-to-thickness ratios and weight fractions of GOs ($L/h = 25$, $K_w = 100$, $K_p = 10$).

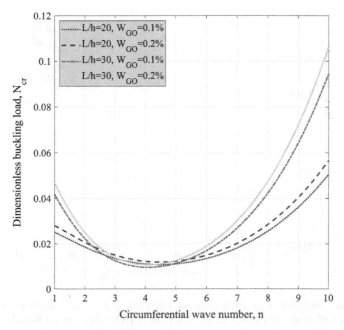

FIGURE 3.5 Variation of the dimensionless buckling load of S-S nanocomposite shells versus circumferential wave number for different length-to-thickness ratios and weight fractions of GOs ($R/h = 25$, $K_w = 100$, $K_p = 10$).

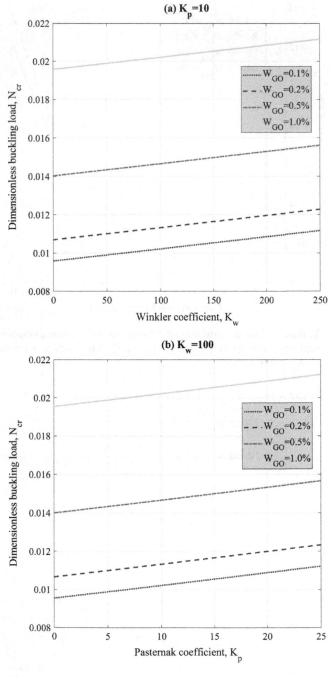

FIGURE 3.6 Variation of dimensionless buckling load of S-S nanocomposite shells versus foundation coefficients for various weight fractions of GOs whenever (a) Winkler coefficient is varied and (b) Pasternak coefficient is varied ($L/h = R/h = 20$, $n = 4$).

In Figure 3.6, the fundamental buckling load of the GOR nanocomposite shell (i.e., the buckling load in $n = 4$) is examined to analyze the qualitative impacts of the foundation parameters on the stability endurance of the shell-type element. It can be seen that once either the Winkler or Pasternak coefficient is fixed, the shell can tolerate greater buckling loads when the unfixed coefficient of the elastic medium is added. So, both Winkler and Pasternak coefficients possess amplifying impacts on the buckling load that can be endured by the nanocomposite shell. In other words, increasing the stiffness of the elastic medium results in an enhancement in the equivalent stiffness of the continuous system, which leads to an increase in the buckling load of the system. In addition, it must be noticed that the effect of Pasternak coefficient is more powerful than that of the Winkler coefficient. As can be seen in this figure, identical amplification can be induced in the buckling load of the nanocomposite shell by changing the Winkler coefficient from 0 to 250 or by changing the Pasternak coefficient from 0 to 25.

As the last illustration in this chapter, Figure 3.7 shows the combined impacts of the GOs' weight fraction and radius-to-thickness ratio of the shell on the buckling behaviors of GOR nanocomposite shells for a slender shell with length-to-thickness ratio of one hundred. It can be seen that the fundamental buckling load of the shell will go through an increasing path as the weight fraction of the GOs grows. This fact was observed in previous diagrams, too. On the other hand, a gradual decrease can be reported for the buckling load of the shell whenever the radius-to-thickness ratio

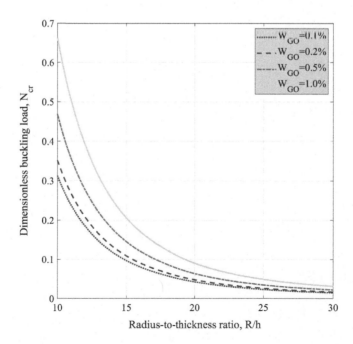

FIGURE 3.7 Variation of the dimensionless buckling load of GOR nanocomposite shells versus radius-to-thickness ratio for various contents of GOs' weight fraction ($L/h = 100$, $n = 4$, $K_w = 100$, $K_p = 10$).

is increased. The major reason for this phenomenon is the transformation of the shell to a thinner one as the radius-to-thickness ratio grows. Hence, the flexibility of the shell will be increased, which results in a reduction in the buckling load of the shell due to the fact that increasing the system's flexibility decreases its stiffness.

3.3 CONCLUDING REMARKS

In this chapter, the effects of various contents of GOs on the buckling behaviors of GOR nanocomposite cylinders were included in an analytical investigation founded on the basis of the well-known FSDT of shells. It was demonstrated that the buckling endurance of the cylindrical shell can be increased by employing greater amounts of GOs in the composition of the constituent nanocomposite material. It was also shown that using high coefficients for the elastic springs of the elastic substrate reveals greater stability bounds for the nanocomposite structure. On the other hand, using high radius-to-thickness ratios causes a decrease in the buckling loads that can be applied to the nanocomposite structure without reaching the buckled state.

4 Thermally Influenced Stability Behaviors of GOR Nanocomposite Cylinders

4.1 PROBLEM DEFINITION

This chapter will examine the buckling problem of an FG-GOR nanocomposite single-layered shell whenever the system is subjected to thermal loading. The shell is assumed to be embedded on an elastic medium and its geometry and the configuration system required to formulate the buckling problem can be observed in Figure 3.1, presented in the last chapter. The governing equations of motion are presented in Chapter 2, Eqs. (2.77)–(2.81). This study includes the effects of various temperature distributions on the stability responses of the nanocomposite cylinder. In other words, different types of temperature profiles, including uniform, linear, and sinusoidal, are included to determine that which one affects the stability responses of the cylinder more than the others. The temperature rises considered in this study can be presented in the following form:

$$T = T_0 + \Delta T \qquad \text{for uniform temperature rise (UTR)} \qquad (4.1)$$

$$T = T_0 + \Delta T \left(\frac{z}{h} + \frac{1}{2} \right) \qquad \text{for linear temperature rise (LTR)} \qquad (4.2)$$

$$T = T_0 + \Delta T \left[1 - \cos\left(\frac{\pi z}{2h} + \frac{\pi}{4} \right) \right] \qquad \text{for sinusoidal temperature rise (STR)} \quad (4.3)$$

in which T_0 stands for the reference temperature or the room temperature. The temperature rise, which is measured in terms of Kelvin, is shown with ΔT.

4.2 NUMERICAL RESULTS AND DISCUSSION

Graphical illustrations will be employed to discuss the impacts of various types of thermal environment on the stability characteristics of FG-GOR nanocomposite shells rested on an elastic medium. In this chapter, the same material properties as the previous chapter are implemented for the GO and the polymeric matrix. The structure under observation is a single-layered cylindrical shell that is reinforced

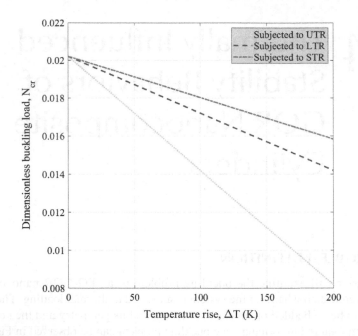

FIGURE 4.1 Variation of the dimensionless buckling load of S-S nanocomposite shells versus temperature rise for various types of thermal loading ($R/h = 20$, $L/h = 20$, $W_{GO} = 0.01$, $K_w = 100$, $K_p = 10$, $n = 4$).

with FG distribution of the reinforcing GOs across the thickness direction. As shown in the previous chapter, the fundamental buckling load of the nanocomposite shell can be obtained using $n = 4$. Therefore, this buckling response will be utilized to estimate the stability behaviors of the FG-FOR nanocomposite shell under various types of thermal loading.

Figure 4.1 reveals the variation of the fundamental buckling load of the UD-GOR nanocomposite shells against temperature rise for various types of temperature distributions applied on the continuous system. It can be seen that the buckling load of the shell decreases continuously as the temperature rise grows. Indeed, the equivalent stiffness of the constituent material reduces in the case of using a remarkable thermal gradient. Thus, it is natural to see such a reduction in the buckling load of the structure. It is interesting to point out that among all types of thermal forces, the worst condition belongs to the case of subjecting the nanocomposite to a uniformly distributed temperature rise across the thickness of the system. In such case, all parts of the shell are placed under a uniform temperature rise; whereas in linear-type or sinusoidal-type temperature rises, the magnitude of the temperature rise depends on the distance from the neutral surface of the shell which leads to a fluctuation in the strength of the applied thermal force. As a brief conclusion, the UTR must be included in any stability analysis dealing with designing a nanocomposite shell reinforced with GOs.

Next, Figure 4.2 illustrates the effect of thermal gradients on the variation of the dimensionless buckling load of the GOR nanocomposite shells versus circumferential

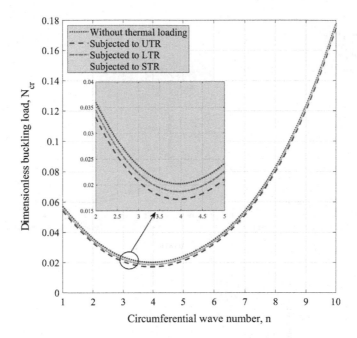

FIGURE 4.2 Variation of the dimensionless buckling load of S-S nanocomposite shells versus circumferential wave number for various types of thermal loading ($R/h = 20$, $L/h = 20$, $W_{GO} = 0.01$, $Kw = 100$, $Kp = 10$, $\Delta T = 50$ K).

wave number once the reinforcing GOs are assumed to be distributed across the shell's thickness in a uniform manner. It can be seen that the stability limit of the cylindrical shell will shift downward once a thermally influenced environment is utilized. This trend is an outcome of the stiffness softening phenomenon appeared in the nanocomposite material due to the increase of the temperature. It is not strange to observe a decrease in the buckling load of the shell in this case. On the other hand, the results of previous figure can be reinforced, in that the worst stability circumstance relates to the case of subjecting the nanocomposite shell to a uniform-type thermal loading compared with linear and sinusoidal ones. It must also be considered that the buckling response of the nanocomposite structure will be influenced by thermal loading in the mid-range circumferential wave numbers more than very small or very big ones.

Figure 4.3 shows the influence of various types of temperature rises on the buckling load of the nanocomposite shells once the circumferential wave number is varied from 1 to 10 with respect to different contents of reinforcing GOs in the composition of the constituent material. On the basis of this figure, the GOR nanocomposite shell will be able to tolerate greater buckling excitations in the case of implementing greater contents of GOs in the nanocomposite. Indeed, using greater amount of GOs results in an improvement in the stiffness of the nanocomposite material which causes an amplification in the cross-sectional rigidities of the shell. Thereafter, the equivalent stiffness of the nanocomposite

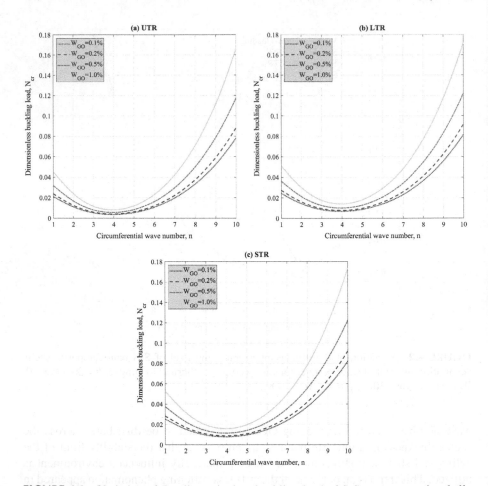

FIGURE 4.3 Variation of the dimensionless buckling load of S-S nanocomposite shells versus circumferential wave number for various weight fractions of GOs for **(a)** UTR, **(b)** LTR, and **(c)** STR ($R/h = 20$, $L/h = 20$, $K_w = 100$, $K_p = 10$, $\Delta T = 200$ K).

structure will grow and this improvement results in a higher stability limit for the continuous system. According to this diagram, it can be seen that among all types of thermal loading, the UTR-type can affect the buckling behaviors of the structure more than the other types. Hence, this type of thermal loading must be addressed in the design procedures. It is also interesting to point out that the most remarkable change in the buckling load can be reported for circumferential wave numbers either smaller than 2 or greater than 9.

In Figure 4.4, the focus is on investigating the effect of employing various types of FG nanocomposites on the determination of the buckling load of the nanocomposite cylinder once the structure is subjected to a UTR-type thermal loading. The buckling responses of the shell can be dramatically affected by the type of distribution

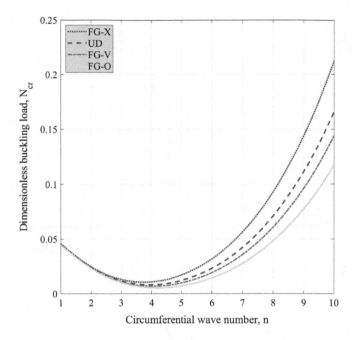

FIGURE 4.4 Variation of the dimensionless buckling load of S-S nanocomposite shells versus circumferential wave number for various distribution of GOs in the media ($R/h = 20$, $L/h = 20$, $W_{GO} = 1\%$, $K_w = 100$, $K_p = 10$, $\Delta T = 200$ K).

that is selected for the reinforcement of the material. However, the intensity of this effect differs as various amounts are assigned to the circumferential wave number. Actually, each of the UD or FG nanocomposites can be employed in small circumferential wave numbers without any particular difference in the stability limit of the structure; however, selecting FG nanocomposites can either strengthen or lessen the endurable buckling load of the structure in circumferential wave numbers greater than 3. In fact, FG-X nanocomposite shells reinforced with GOs can provide greater buckling loads than those tolerated by UD nanocomposites. However, FG-O nanocomposites are weaker than UD nanocomposites in enduring buckling loads. The reason for this is that in the FG-X nanocomposites, the surfaces with longer distance from the neutral surface of the shell (i.e., surfaces with greater radius for calculation of the stress resultants) possess a greater content of GOs and those closer to the neutral surface, who are not involved in determination of the stress resultants as important as previous ones, have a lower content of reinforcing GOs. This is the reason for the better performance of FG-X nanocomposites in comparison with UD ones. However, the reverse pattern is valid for the FG-O nanocomposites and due to this fact, these nanocomposites are weaker than UD ones. So, it is of significance to select proper distribution pattern for the GOs to produce a cylinder that is able to endure great buckling loads.

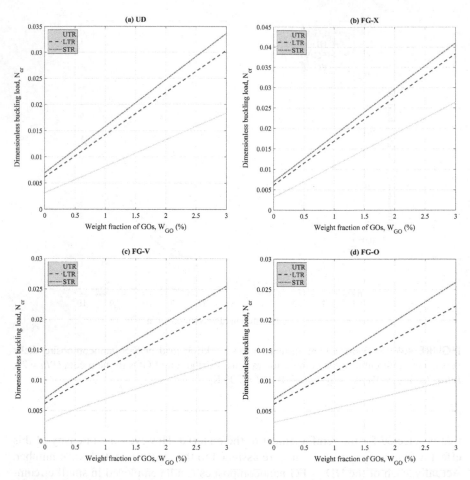

FIGURE 4.5 Variation of the fundamental buckling load of the GOR nanocomposite cylindrical shells against weight fraction of GOs for various types of temperature rise for **(a)** UD, **(b)** FG-X, **(c)** FG-V, and **(d)** FG-O nanocomposites ($L/h = 20$, $R/h = 20$, $K_w = 100$, $K_p = 10$, $n = 4$).

The effect of FG distribution patterns of GOs on the buckling characteristics of GOR nanocomposite shells is again illustrated in Figure 4.5 by plotting the variation of the fundamental buckling load of the cylindrical shells against weight fraction of the GOs for three types of thermal loading. Based upon this diagram, the greater the content of the available GO in the media, the greater will be the stiffness of the system and the stability of the structure will be enhanced. Also, in all types of nanocomposite materials, the buckling load of the shell can be affected by the UTR-type thermal loading more than other types. This trend was observed in previous illustrations, too. Again, note that FG-X GOR nanocomposite shells are the best alternative for the cases where the cylinder is going to be designed for being subjected to a great buckling load.

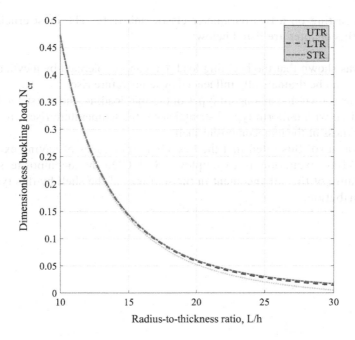

FIGURE 4.6 Variation of the fundamental buckling load of the GOR nanocomposite with FG-O distribution cylinders versus radius-to-thickness ratios for three types of temperature rises applied on the structure ($L/h = 100$, $W_{GO} = 1\%$, $K_w = 200$, $K_p = 20$, $n = 4$).

The final illustration in this chapter is dedicated to understanding the effect of the radius-to-thickness ratio of the shell on the fundamental buckling load of the FG-O nanocomposite structure whenever the nanocomposite is subjected to various types of thermal loading (see Figure 4.6). As predicted before, the worst thermal working conditions for the nanocomposite shell occur once the shell is placed in a thermal environment with UTR. The influence of the type of temperature rise on the buckling behavior of the nanocomposite shell can be better seen in big radius-to-thickness ratios. It is also observable that the buckling load of the shell will decrease in the case of increasing the ratio between the radius and the thickness of the nanocomposite structure. In the previous chapter, it was explained that increasing the radius-to-thickness ratio results in a thinner structure with greater flexibility. Therefore, it is natural to see a reduction in the buckling load of the shell in such a case due to the inverse relation between the flexibility and the stiffness in linear elastic solids.

4.3 CONCLUDING REMARKS

The goal of this chapter was to emphasize the crucial role of FG patterns of reinforcing GOR nanocomposite materials and the existence of thermal gradients of different types on the stability characteristics of FG-GOR nanocomposite cylinders resting on a two-parameter elastic substrate. The most crucial high-lights of this chapter are listed below:

- It was shown that the buckling load that can be tolerated by a cylindrical shell can be dramatically influenced by temperature rise.
- It was shown that among all types of thermal loading, the worst one is the loading with uniform-type distribution for the temperature rise across the thickness of the nanocomposite shell.
- It was also illustrated that the best choice for cases of having extreme buckling excitation is to employ FG-O GOR nanocomposite shells because of the enhancement in the stiffness of the shell by this type of distribution.

5 Thermal Buckling Behaviors of Sandwich Beams Covered by FG-CNTR Nanocomposite Facesheets

5.1 PROBLEM DEFINITION

This chapter will examine the buckling problem of a three-layered sandwich beam in the presence of thermal gradient. The structure under observation is a sandwich beam with a homogeneous core surrounded by two FG nanocomposites reinforced with CNTs. The dependency of the constituent materials on the local temperature is covered using temperature-dependent material properties for the homogeneous core and nanocomposite facesheets. The schematic view of the beam-type structure studied in this chapter can be observed in Figure 5.1.

The equivalent material properties of the nanocomposite facesheets can be obtained using the relations provided in Section 2.3.1, Eq. (2.234). The governing differential equations of the problem were generated in the framework of the principle of virtual work incorporated with the FSDT of beams, known as Timoshenko theory. The governing equations can be seen in Chapter 2, Eqs. (2.112)–(2.114) by ignoring the terms that include time derivatives. It must be mentioned that in this chapter, due to the changes in the components of the stiffness matrix of the material across the thickness of the beam, the integrations over the thickness of the beam (i.e., integrations from $z = -h/2$ to $z = h/2$) must be discretized to three integrals as follows:

$$\int_{-\frac{h}{2}}^{\frac{h}{2}} F(z)dz = \int_{-\frac{h}{2}}^{-\frac{h}{2}+h_f} F(z)dz + \int_{-\frac{h}{2}+h_f}^{\frac{h}{2}-h_f} F(z)dz + \int_{\frac{h}{2}-h_f}^{\frac{h}{2}} F(z)dz \qquad (5.1)$$

where $F(z)$ stands for any desired integrand which may be either the stiffness resultant or the inertia resultant. Finally, it is important to point out that the governing equations will be solved according to the well-known DTM as a powerful and

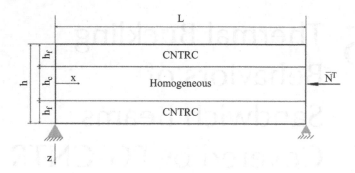

FIGURE 5.1 The geometry of the sandwich beam with homogeneous core and FG-CNTR nanocomposite facesheets.

rapid numerical method to develop the buckling load of the sandwich beam for both clamped-type and simply supported-type BCs. The instructions for how to utilize this method in solving the static or dynamic problems of a Timoshenko beam via this method can be found in Chapter 2, Section 2.2.3.

5.2 NUMERICAL RESULTS AND DISCUSSION

In this section numerical examples will be presented to study the effects of different variants on the buckling characteristics of beam-type structures. Before beginning the numerical investigation the constituent materials must be introduced. The core of the sandwich beam is manufactured from $Ti - 6Al - 4V$ with material properties tabulated in Table 5.1.

Based on the data provided in Table 5.1 for the temperature-dependent material properties of the core, the Young's modulus and the CTE of the core at any desired temperature can be calculated using the following formulas:

$$
\begin{aligned}
E &= E_0 \left(E_{-1} T^{-1} + 1 + E_1 T + E_2 T^2 + E_3 T^3 \right), \\
\alpha &= \alpha_0 \left(\alpha_{-1} T^{-1} + 1 + \alpha_1 T + \alpha_2 T^2 + \alpha_3 T^3 \right)
\end{aligned}
\tag{5.2}
$$

where T shows the local temperature. Moreover, the material properties of the implemented CNTs are provided in Table 5.2 for three local temperatures, namely 300, 400, and 500 Kelvin. Also, the size-dependent coefficients of the CNTs, used

---

TABLE 5.1

Temperature-Dependent Properties of Young's Modulus and CTE for Ti-6Al-4V

Properties	P_0	P_{-1}	P_1	P_2	P_3
$E\ (Pa)$	122.56e9	0	-4.586e-4	0	0
$\alpha\ (1/K)$	7.5788e-6	0	6.638e-4	-3.147e-6	0

TABLE 5.2
Temperature-Dependent Material Properties of CNTs

Temperature (K)	E_{11} (TPa)	E_{22} (TPa)	G_{12} (TPa)	α (1 / K)
300	5.6466	7.0800	1.9445	3.4584
500	5.5308	6.9348	1.9643	4.5361
700	5.4744	6.8641	1.9644	4.6677

in Eq. (5.2), can be assumed to be as same as those reported by Han and Elliott (2007). Also, the material properties of the PMMA polymeric matrix which is used to host the CNTs in the nanocomposite facesheets can be obtained using the data reported by Yang et al. (2015).

Now the numerical results can be presented to analyze the buckling characteristics of sandwich beams. In Table 5.3 the convergence study is fulfilled to show the number of iterations required to obtain the accurate buckling load of the structure. It is shown that the buckling load of the beam converges after 19 iterations with the criterion of 4-digit precision. Therefore, 19 iterations will be utilized in all of the future studies.

TABLE 5.3
The Influence of Number of Iterations on the Convergence of the Buckling Load of the Beam Using the DTM

N	N_{cr}
6	-
7	-
8	-
9	-
10	-
11	0.02975
12	0.02243
13	0.02425
14	0.02371
15	0.02723
16	0.02690
17	0.02583
18	0.02583
19	0.02591
20	0.02591
21	0.02591
22	0.02591
23	0.02591

TABLE 5.4

Comparison of the Dimensionless Critical Buckling Load of the Sandwich Beams with FG-CNTR Nanocomposite Facesheets ($h_c/h_f = 8$)

Simply Supported BC at Both Ends

		$V^*_{CNT} = 0.12$		$V^*_{CNT} = 0.17$		$V^*_{CNT} = 0.28$	
L/h	Type	Present	Wu et al. (2015)	Present	Wu et al. (2015)	Present	Wu et al. (2015)
10	FG	0.0071	0.0072	0.0086	0.0085	0.0115	0.0111
	UD	0.0069	0.0070	0.0083	0.0082	0.0110	0.0107
20	FG	0.0018	0.0018	0.0022	0.0022	0.0029	0.0029
	UD	0.0018	0.0018	0.0021	0.0021	0.0028	0.0028
30	FG	0.0008	0.0008	0.0010	0.0010	0.0013	0.0013
	UD	0.0008	0.0008	0.0009	0.0009	0.0012	0.0012

Clamped BC at Both Ends

		$V^*_{CNT} = 0.12$		$V^*_{CNT} = 0.17$		$V^*_{CNT} = 0.28$	
L/h	Type	Present	Wu et al. (2015)	Present	Wu et al. (2015)	Present	Wu et al. (2015)
10	FG	0.0259	0.0261	0.0309	0.0305	0.0408	0.0387
	UD	0.0252	0.0254	0.0305	0.0296	0.0400	0.0373
20	FG	0.0071	0.0072	0.0088	0.0085	0.0113	0.0111
	UD	0.0069	0.0070	0.0083	0.0082	0.0110	0.0107
30	FG	0.0032	0.0032	0.0039	0.0039	0.0051	0.0051
	UD	0.0031	0.0037	0.0038	0.0037	0.0050	0.0049

Another tabular study is conducted to show the reliability of the presented methodology by setting a comparison study between the results reported by Wu et al. (2015) and those developed using the present model. Based on this comparison study, tabulated in Table 5.4, the buckling loads of the sandwich beams with both UD and FG nanocomposites reinforced with CNTs are compared. It can be seen that for both simply supported and clamped BCs, the buckling loads of present modeling are in an excellent agreement with those reported in the open literature.

Table 5.5 shows the results of the investigation of the coupled effects of the CNTs' volume fraction and type of the implemented CNTR nanocomposite in the facesheets of the sandwich beam on the buckling load of the structure. Clearly, the stability limit of the structure can be improved by adding the volume fraction of the CNTs. It was discussed in previous chapters that the stiffness of the nanocomposite can be enhanced by adding the content of the nanosize reinforcements inside it. However, FG nanocomposites provide better buckling loads than the UD ones. The reason for this is that in this study, FG-X nanocomposites were utilized, which can improve the equivalent stiffness of the nanocomposite material more

TABLE 5.5

Effect of Nanotube Volume Fraction and Slenderness Ratio on Dimensionless Critical Buckling of Sandwich Beams with FG-CNTRC Facesheets ($h_c/h_f = 8$)

			$\Delta T = 0$			$\Delta T = 200$			$\Delta T = 400$		
			V_{CNT}^*			V_{CNT}^*			V_{CNT}^*		
L/h	BC	Type	0.12	0.17	0.28	0.12	0.17	0.28	0.12	0.17	0.28
10	S-S	FG	0.0071	0.0086	0.0115	0.0066	0.0082	0.0109	0.0062	0.0076	0.0104
	S-S	UD	0.0069	0.0083	0.0110	0.0064	0.0079	0.0105	0.0060	0.0073	0.0099
20	S-S	FG	0.0018	0.0022	0.0029	0.0017	0.0020	0.0028	0.0016	0.0019	0.0026
	S-S	UD	0.0018	0.0021	0.0028	0.0017	0.0018	0.0027	0.0015	0.0018	0.0025
30	S-S	FG	0.0018	0.0010	0.0013	0.0007	0.0009	0.0012	0.0007	0.0008	0.0011
	S-S	UD	0.0018	0.0009	0.0012	0.0007	0.0008	0.0011	0.0007	0.0008	0.0011
10	C-C	FG	0.0259	0.0309	0.0408	0.0240	0.0290	0.0376	0.0227	0.0276	0.0373
	C-C	UD	0.0252	0.0305	0.0400	0.0234	0.0281	0.0367	0.0226	0.0266	0.0365
20	C-C	FG	0.0071	0.0088	0.0113	0.0065	0.0079	0.0105	0.0062	0.0076	0.0103
	C-C	UD	0.0069	0.0083	0.0110	0.0063	0.0077	0.0103	0.0061	0.0072	0.0101
30	C-C	FG	0.0032	0.0039	0.0051	0.0029	0.0036	0.0046	0.0028	0.0034	0.0047
	C-C	UD	0.0031	0.0038	0.0050	0.0028	0.0034	0.0045	0.0027	0.0033	0.0046

than the simple UD nanocomposites. Also, adding the slenderness ratio of the beam (i.e., the length-to-thickness ratio) results in an increase of the material's flexibility which corresponds with lower stiffness. Hence, it is obvious that the buckling load that can be tolerated by the structure becomes smaller in the case of increasing this ratio. In addition, it is very clear that the stability endurance of the sandwich structure becomes smaller once a nonzero value is assigned to the temperature rise. Indeed, increasing the temperature from its reference value results in a reduction in the stiffness of the material and makes the structure softer. Therefore, it can be observed that the buckling load of the beam is decreased in the case of increasing the temperature rise.

The influences of different types of BCs, the core-to-facesheet thickness ratio, the volume fraction of the CNTs, and the thermal environment on the buckling responses of sandwich beams are included in Figure 5.2, plotting the variation of the buckling load against volume fraction of the CNTs in the nanocomposite facesheets. It can be seen that the beams with a greater core-to-facesheet thickness ratio can tolerate smaller buckling loads. Indeed, the stiffness of the structure decreases as the volume fraction of the core phase grows in the composition of the structure. Also, C-C beams are better candidates for situations where the structure is assumed to be subjected to a huge static excitation. Again, it can be determined that increasing the temperature rise applied on the structure reveals lower buckling responses. This was observed in the previous diagram, too, and the reason for this phenomenon was explained in detail.

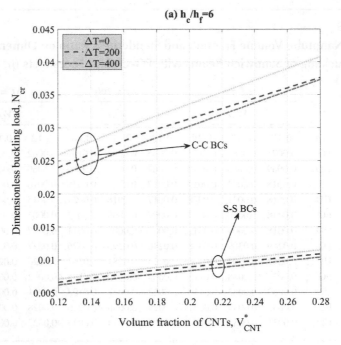

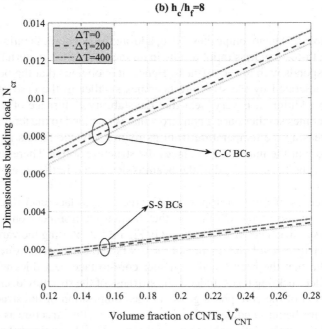

FIGURE 5.2 Variation of the dimensionless buckling load of the sandwich beams versus volume fraction of the CNTs for both S-S and C-C BCs once (a) $h_c/h_f = 6$ and (b) $h_c/h_f = 8$.

TABLE 5.6
Effect of Nanotubes' Volume Fraction and h_c/h_f on the Dimensionless Critical Buckling Load of Sandwich Beams with FG-CNTR Nanocomposite Facesheets ($L/h = 20$)

			$\Delta T = 0$			$\Delta T = 200$			$\Delta T = 400$		
			V^*_{CNT}			V^*_{CNT}			V^*_{CNT}		
h_c/h_f	BC	Type	0.12	0.17	0.28	0.12	0.17	0.28	0.12	0.17	0.28
8	S-S	FG	0.0018	0.0022	0.0029	0.0017	0.0020	0.0028	0.0016	0.0019	0.0026
	S-S	UD	0.0018	0.0021	0.0028	0.0016	0.0018	0.0027	0.0015	0.0018	0.0025
6	S-S	FG	0.0018	0.0023	0.0032	0.0017	0.0021	0.0030	0.0016	0.0020	0.0029
	S-S	UD	0.0017	0.0022	0.0030	0.0016	0.0020	0.0028	0.0015	0.0019	0.0027
4	S-S	FG	0.0019	0.0024	0.0036	0.0017	0.0023	0.0034	0.0016	0.0022	0.0033
	S-S	UD	0.0017	0.0022	0.0033	0.0016	0.0021	0.0031	0.0015	0.0020	0.0029
8	C-C	FG	0.0071	0.0088	0.0113	0.0065	0.0079	0.0105	0.0062	0.0076	0.0103
	C-C	UD	0.0069	0.0083	0.0110	0.0063	0.0077	0.0103	0.0061	0.0072	0.0101
6	C-C	FG	0.0071	0.0088	0.0122	0.0066	0.0083	0.0117	0.0063	0.0080	0.0114
	C-C	UD	0.0067	0.0083	0.0115	0.0063	0.0079	0.0110	0.0060	0.0076	0.0107
4	C-C	FG	0.0071	0.0093	0.0136	0.0068	0.0089	0.0131	0.0065	0.0087	0.0129
	C-C	UD	0.0066	0.0086	0.0124	0.0064	0.0082	0.0119	0.0060	0.0079	0.0117

Furthermore, the effects of the core-to-facesheet thickness ratio and volume fraction of the CNTs on the critical buckling load of the sandwich beam is tabulated in Table 5.6. It is clear that implementing higher temperature rises can result in a reduction in the buckling load of the sandwich beam. As explained before, the reason for this phenomenon is the softening behavior intensified in the sandwich structure by increasing the temperature. Also, it is evident that the buckling load of the beam-type structure goes through a decreasing path as the core-to-facesheet thickness ratio becomes greater. Indeed, the volume fraction of the nanocomposite facesheets in the structure is reduced, which leads to a reduction in the equivalent stiffness of the sandwich element. Clearly, the beams with C-C BCs are more stable than those with S-S BCs as predicted before.

5.3   CONCLUDING REMARKS

This chapter of the book examined the influence of increasing the local temperature on the stability characteristics of sandwich beams surrounded by two nano-composite facesheets. The problem's formulation was based on the expansion of the principle of virtual work for a Timoshenko beam theory. The homogenization of the nanocomposite facesheets was fulfilled using the micromechanical methods for the CNTR nanocomposite materials considering the effect of FG distribution of the nanofillers across the thickness of the facesheets. The highlights of this chapter can be stated as below:

- It was concluded that the FG-O distribution of the CNTs can help the designer to apply larger buckling loads on the structure.
- It was shown that the stability of the beam can be highly affected by increasing the temperature rise from zero to any other nonnegative value.
- Results of present chapter revealed that sandwich beams with smaller core-to-facesheet thickness ratios can be better alternatives for the cases where a great excitation is going to be applied to the designed element.

6 Thermo-Magnetically Influenced Static Stability of FG-GOR Nanocomposite Beams

6.1 PROBLEM DEFINITION

This chapter examines the influence of a thermal gradient on the stability behaviors of FG-GOR nanocomposite beams as well as the influence of a non-uniform external magnetic field that exists in the environment. The derivation of the governing equations of the problem will be completed using the principle of virtual work in association with the kinematic relations of the refined higher-order beam theory. The governing equations of the problem can be extracted from Eqs. (2.166)–(2.168), dismissing the dynamic terms. In fact, the terms including differentiation with respect to time must be excluded from the governing equations to develop the governing equations of the buckling problem for the defined working condition. Afterward, the derived governing equations must be solved by an analytical solution to derive the buckling load of the beam-type element. In the present chapter, the beam will be considered to be a single-layered one filled with GOR nanocomposite material that may be either UD or FG. In what follows, numerical illustrations will be provided for the goal of highlighting the effects of different variants on the buckling behaviors of the polymeric nanocomposite beams.

6.2 NUMERICAL RESULTS AND DISCUSSION

In this section, a series of schematic studies are carried out to investigate the effects of various parameters such as temperature change, GOs' weight fraction, magnetic field intensity, different types of GOs' distribution, and various BCs on the buckling behavior of GOR nanocomposite beams. The effective material properties of the GOR nanocomposite beams are chosen to be the same as those reported by Zhang et al. (2018). The GOs possess negative CTE and the amount of this coefficient is considered to be -50^{e-6} $(1/K)$ based on earlier research (Su et al., 2012). At the beginning of this section, the accuracy of the present model is verified. The natural frequencies of S-S CNTR nanocomposite beams are obtained by the present model and compared with those of Wattanasakulpong and Ungbhakorn (2013) and Zhang et al. (2018). The results of the verification can be observed in tabulated form in Table 6.1 and Table 6.2.

TABLE 6.1

Comparison of the First Dimensionless Frequency of S-S CNTR Nanocomposite Beams ($L/h = 15$, $V^*_{CNT} = 0.12$)

	Distribution type		
Reference	UD	FG-O	FG-X
Wattanasakulpong and Ungbhakorn (2013)	0.9976	0.7628	1.1485
Zhang et al. (2018)	0.9842	0.7595	1.1249
Present model	0.9904	0.7528	1.1399

In all of the future results, the following dimensionless forms of buckling load and magnetic field intensity are used:

$$N_{cr} = \frac{N^b L^2}{E_{GO} I}, \qquad H_0 = \frac{\eta A L^2}{E_{GO} I} \bar{H}_x^2, \qquad I = \frac{b h^3}{12} \qquad (6.1)$$

Figure 6.1 considers the effect of changing the temperature rise on the stability characteristics of different types of FG-GOR nanocomposite beams. To this purpose, the variation of the dimensionless buckling load of the nanocomposite beams was plotted for the first five mode numbers in the presence of an external magnetic field for the case of S-S beams. It can be seen that the buckling endurance of the structure becomes greater in higher modes. In addition, it is clear that among all types of FG nanocomposites, FG-X nanocomposites can tolerate higher buckling loads in comparison with other types of nanocomposites. The reason for this is that the equivalent material properties of the beam will be greater in the spatial coordinates that possess greater distance from the neutral axis of the beam, which leads to an increase in the cross-sectional rigidities of the beam. On the other hand, this figure shows that increasing the temperature rise applied to the beam results in a decreasing effect on

TABLE 6.2

Comparison of the First Dimensionless Frequency of S-S GOR Nanocomposite Beams ($W_{GO} = 0.3\%$)

L/h		Distribution type		
		FG-X	FG-O	UD
10	Zhang et al. (2018)	0.3379	0.2921	0.3159
	Present model	0.3576	0.3013	0.3095
15	Zhang et al. (2018)	0.2271	0.1959	0.2121
	Present model	0.2411	0.2009	0.2079
20	Zhang et al. (2018)	0.1708	0.1473	0.1595
	Present model	0.1815	0.1461	0.1564

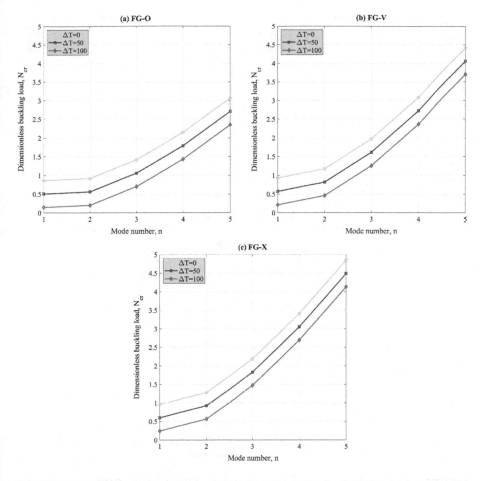

FIGURE 6.1 Variation of the dimensionless buckling load of a S-S GOR nanocomposite beam versus mode number $\left(L/h = 25, W_{GO} = 1\%, \bar{H}_x = 0.75 \right)$.

the stiffness of the beam and this negative effect causes a reduction in the stability load of the beam. So, it must be regarded by designers that implementation of FG-O nanocomposites in thermal environment can be very dangerous if a large buckling excitation is going to be applied to the beam-type element.

Investigating the combined influences of BCs and distribution patterns of GOs across the thickness of the beam on the fundamental buckling load of the nanocomposite beams is covered in the framework of Figure 6.2. According to this figure, it can be seen that in any desired temperature rise, the FG-X nanocomposites possess the greatest response followed by FG-V and FG-O ones. It is also clear that the softening impact of the temperature rise on the equivalent stiffness of the nanocomposite results in a gradual decrease in the buckling load of the beam. Moreover, like our previous estimates, the beams with clamped-type BCs can endure larger buckling loads compared with those with simply supported ones. This is because

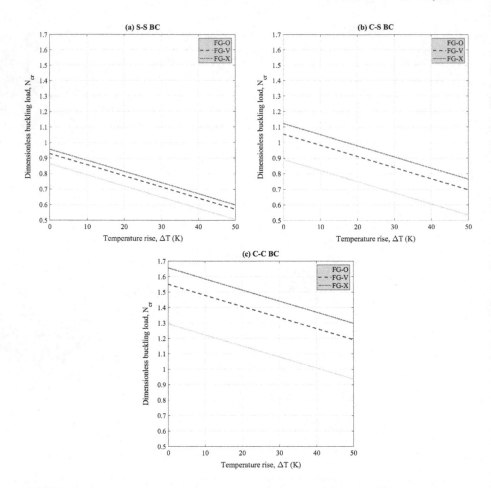

FIGURE 6.2 Variation of the first dimensionless buckling load of a GOR nanocomposite beam versus temperature rise for various types of BCs regarding different distribution patterns of GOs across the thickness $\left(L/h = 25, W_{GO} = 1\%, \bar{H}_x = 0.75\right)$.

the structural stiffness of the beam will be increased in the case of implementing a clamped BC rather than a simply supported one.

Emphasizing the important effect of the magnetic field on the static stability of the FG nanocomposite beams, Figure 6.3 depicts the variation of the fundamental buckling load of the nanocomposite beams versus magnetic field intensity for various types of FG nanocomposites and either existence or absence of thermal gradient in the environment. It is obvious that the buckling limit of the nanocomposite beam can be amplified by using FG-X nanocomposites. Also, an increase in the intensity of the magnetic field results in an amplification in the stability response of the structure because the induced magnetic force affects the continuous system in a way that results in an increase in its equivalent stiffness matrix. As with previous results, the beams with C-C BCs can survive under greater buckling loads compared with C-S and S-S beams.

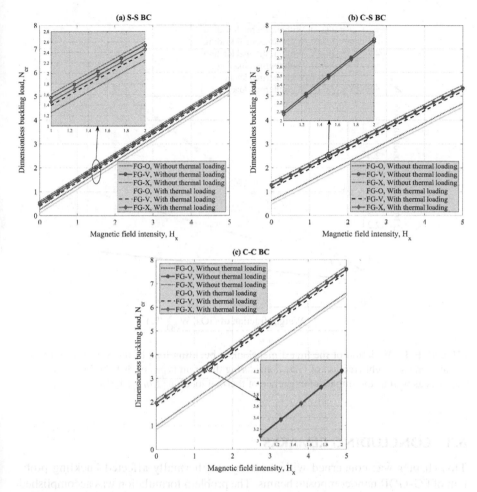

FIGURE 6.3 Coupled effects of magnetic field intensity, temperature rise, different distributions of the GOs across the thickness, and various types of BCs on the fundamental buckling responses of FG-GOR nanocomposite beams $\left(L/h = 10, W_{GO} = 4\% \right)$.

The final illustration, Figure 6.4, plots the variation of the fundamental buckling load of the beams consisting of FG nanocomposites against weight fraction of GOs once the effects of thermal loading are included or not. Clearly, the buckling load of the nanocomposite structure grows continuously as the weight fraction of the GOs in the nanocomposite material becomes greater. This trend is valid because implementation of a larger amount of reinforcements in the nanocomposite results in an increase in the stiffness of the material, which results in a corresponding increase in the stability limit of the structure. Obviously, FG nanocomposite beams placed in a thermally affected environment can tolerate smaller static loadings compared with those placed in an environment that is not influenced by thermal gradient.

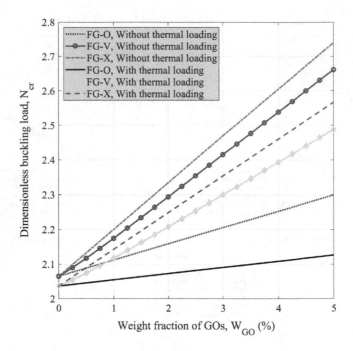

FIGURE 6.4 Variation of the first dimensionless buckling load of the S-S nanocomposite beams versus weight fraction of GOs considering different types of distribution patterns for the GOs as well as the impact of existence of thermal loading $\left(L/h = 10, \bar{H}_x = 2 \right)$.

6.3 CONCLUDING REMARKS

This chapter was concerned with the magneto-thermally affected buckling problem of FG-GOR nanocomposite beams. The problem formulation was accomplished using the refined form of the shear deformable beam hypotheses incorporating the principle of virtual work. The effective material properties of the FG nanocomposite were obtained within the framework of the well-known Halpin-Tsai micromechanical scheme. The influences of various types of edge supports on the mechanical response of the system were included, too. The most crucial results of this study are rewritten below:

- The higher the weight fraction of the GOs in the nanocomposites' composition, the stiffer the achieved nanocomposite and this fact results in an increase in the buckling responses of the nanocomposite structures.
- Assigning a nonzero thermal gradient on the environment results in a decrease in the stability of the beam-type element. Indeed, such thermal loading can result in a reduction in the stiffness of the material and finally affects the buckling behavior of the structure in a decreasing manner.
- It was again noted that FG-X nanocomposites can endure greater excitations compared to other types of FG nanocomposites.

7 Effects of Axial Excitation and Non-Uniform Magnetic Field on the Buckling Behavior of FG-GOR Nanocomposite Beams

7.1 PROBLEM DEFINITION

This chapter examines the simultaneous impacts of the existence of a non-uniform magnetic field in the environment and applying axial loading on the beam-type element on the buckling responses of such a structure consisting of FG-GOR nanocomposite materials, using a refined shear deformable beam model. The equivalent material properties of the nanocomposite can be obtained using the micromechanical method presented in Part 1. Also, the procedure of deriving the governing equations of the problem is as same as that followed in former chapters. However, the time-dependent parts of the governing equations must be ignored due to the fact that the problem under observation is a static one. Hence, Eqs. (2.166)–(2.168) can be employed to formulate the problem. The axial force applied on the beam and the buckling load will be entered in the aforementioned equations just as same as thermal force. However, no thermal gradient exists in the present problem. In what follows, the influences of different parameters on the buckling characteristics of FG-GOR nanocomposite beams will be studied using a set of numerical illustrations.

7.2 NUMERICAL RESULTS AND DISCUSSION

In the present section, the numerical examples related to the static stability behaviors of FG-GOR nanocomposite beams will be presented and an extended discussion will be presented for each of the examples. The material properties of the constituent materials are chosen to be the same as those presented by Zhang et al. (2018). In all of the results, the following dimensionless forms of buckling load and magnetic field intensity are used:

$$N_{cr} = \frac{N^b L^2}{E_{GO} I}, \qquad H_0 = \frac{\eta A L^2}{E_{GO} I} \bar{H}_x^2, \qquad I = \frac{bh^3}{12} \qquad (7.1)$$

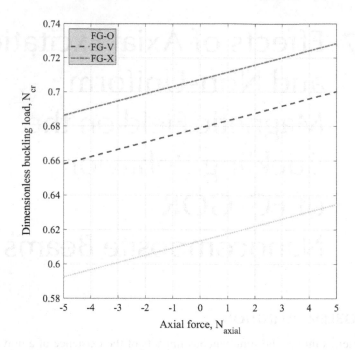

FIGURE 7.1 Variation of the dimensionless buckling load of the FG-GOR nanocomposite beams with S-S BCs versus applied axial force for various types of distribution patterns for the GOs $\left(L / h = 25, K_w = K_p = 0, W_{GO} = 1\%, \bar{H}_x = 0.5\right)$.

In Figure 7.1, the variation of the dimensionless buckling load of the GOR nano-composite beams is drawn against applied axial force for different types of distribution of the reinforcing phase across the thickness of the beam. It can be observed that adding the applied axial force results in a tension in the beam and due to this induced tension, the ability of the beam to endure compressive buckling load will be increased. This trend can be observed in any desired distribution pattern for the GOs. On the other hand, it is clear that the beams manufactured from FG-X nanocomposites can endure higher buckling loads among different types of FG nanocomposites. The reason for this trend was expressed in Part 1 of this book, i.e., the greater cross-sectional rigidities of the beam in the case of using FG-X nanocomposites.

Figure 7.2 displays the coupled impacts of the foundation's stiffness and the magnetic field intensity on the stability curves of UD-GOR nanocomposite beams. It can be found that adding to the stiffness of the foundation's springs can result in an increase in the stability bounds of the nanocomposite structure. In other words, the structural stiffness of the structure will be enhanced, which leads to a greater effective stiffness for the beam-type element. Throughout the direct relationship between the stiffness of the structure and the stability limit, it is natural to observe such an increase in the buckling load of the nanocomposite beam. It is worth mentioning that the influence of the Pasternak springs is many times greater than that of the Winkler ones. In addition, it is clear that the buckling bounds of the system

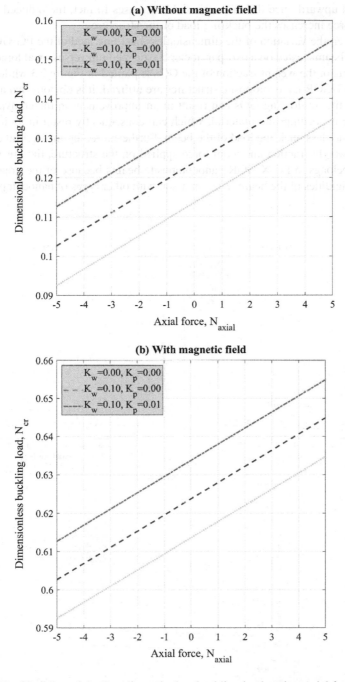

FIGURE 7.2 Variation of the first dimensionless buckling load against axial force for various amounts of foundation parameters considering both existence and absence of external magnetic field $\left(L/h = 25, W_{GO} = 1\% \right)$.

are shifted upward once the magnetic field increases In fact, the induced magnetic Lorentz force increases the buckling load of the structure.

Moreover, the variation of the dimensionless buckling load of the FG-GOR nano-composite beams rested on a two-parameter elastic medium versus axial force for vari-ous amounts of the weight fraction of the GOs is plotted in Figure 7.3 while different types of FG-GOR nanocomposite structures are utilized. It is obvious that adding the content of the reinforcing GOs can result in an improvement in the equivalent stiff-ness of the nanocomposite material, which can consequently result in an increase in the dimensionless buckling load of the beam. Furthermore, for any desired content of the GOs' weight fraction and axial force applied on the structure, the greatest buck-ling load belongs to FG-X GOR nanocomposite beams because of the greater cross-sectional rigidities of the beam in comparison with other types of nanocomposites.

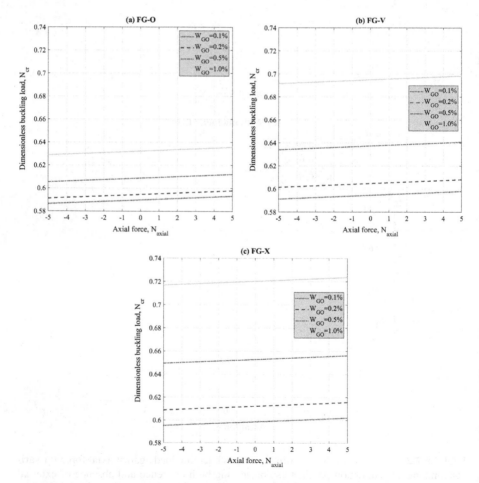

FIGURE 7.3 Variation of the dimensionless buckling load of FG-GOR nanocomposite beams versus axial load for different contents of GOs $\left(L/h = 10, K_w = 0.1, K_p = 0.01, \bar{H}_x = 0.5 \right)$.

As the final numerical example in this chapter, the variation of the dimensionless buckling load of the FG-GOR nanocomposite beam versus weight fraction of the GOs for various types of axial excitation and magnetic field intensity is plotted in Figure 7.4 once different types of distribution patterns of the reinforcing phase are covered. It is clear that in the case of applying traction to the beam, the continuous system will be able to endure higher buckling loads. Also, the buckling response of the FG-GOR nanocomposite structure is increased once the element is placed in a magnetically affected environment. Moreover, the FG-X nanocomposites can show higher stability responses which was perceived in previous illustrations, too.

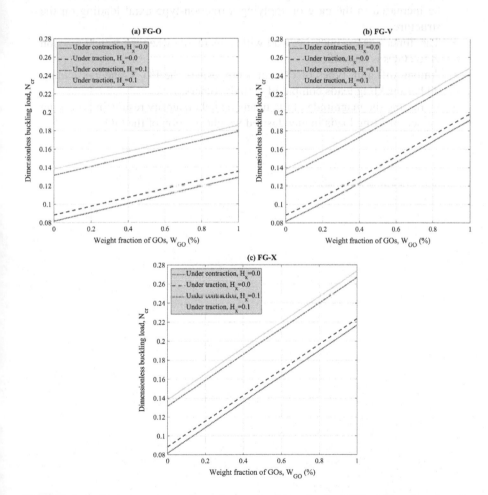

FIGURE 7.4 Variation of the dimensionless buckling load of FG-GOR nanocomposite beams versus weight fraction of GOs for various types of axial loading and different magnetic field intensities $\left(L/h = 10, K_w = 0.1, K_p = 0.01\right)$.

7.3   CONCLUDING REMARKS

This chapter looked at the magneto-elastically influenced buckling responses of FG-GOR nanocomposite beams. The effective material properties of the GOR nanocomposite were obtained using the Halpin-Tsai micromechanical method considering the influence of FG distribution of the reinforcing phase across the thickness of the beam-type element. A non-uniform magnetic field was used to derive the induced Lorentz force applied on the beam. In addition, the kinematic relations were extended on the basis of the refined-type higher-order shear approximation beam models. To put an emphasis on the highlights of this chapter, the most crucial ones will be reviewed here:

- The dimensionless buckling load of the FG-GOR nanocomposite will be increased in the case of applying a traction-type axial loading on the structure.
- The dimensionless buckling load will be increased when the weight fraction of the GOs is intensified.
- Among different types of FG nanocomposites, the FG-X one can endure higher buckling loads compared with the others.
- Increasing the magnitude of the magnetic field intensity results in observing higher buckling loads in any desired weight fraction of the GOs.

8 Thermo-Mechanical Buckling Analysis of Multilayered FG-GPLR Nanocomposite Plates

8.1 PROBLEM DEFINITION

This chapter will probe the effects of thermal gradient on the static stability responses of sandwich multilayered FG-GPLR nanocomposite plates will be probed within the framework of Reddy's plate hypothesis. The effective material properties of the multilayered nanocomposite can be obtained using the instructions presented in Sections 2.3.2 and 2.3.4. Also, the governing differential equations of the problem are similar to those reported in Eqs. (2.191)–(2.195), ignoring the terms related to time due to the static state of the analysis. The governing equations can be solved in an analytical manner for the case of having fully simply supported edges using the well-known Navier's method, as explained in Part 1. In the following sections, the effect of different parameters on the buckling responses of FG-GPLR nanocomposite plates will be studied using some illustrative graphs.

8.2 NUMERICAL RESULTS AND DISCUSSION

This section investigates the influences of various terms on the critical thermal buckling responses of both UD and FG nanocomposite plates reinforced with GPLs. The nanocomposite structure is assumed to be simply supported and is assumed to be subjected to a uniform-type temperature rise. The total number of layers is assumed to be $N_L = 10$, unless another value is reported in the text. Before analyzing the problem under observation, it is best to show that the methodology is powerful enough to predict the buckling responses of plates whenever the structure is subjected to thermal gradient. To this purpose, Table 8.1 compares the critical buckling temperature of SSSS square plates subjected to thermal loading from this study to those reported in the open literature. It can be seen that they are in reliable agreement. The difference between the results of our modeling and those reported by other researchers originates from the fact that in this study the kinematic relations of the plate are written according to a higher-order kinematic hypothesis; whereas the aforementioned references utilized the well-known FSDT to derive the strain-displacement relationships of the plate.

TABLE 8.1

Comparison of the Critical Buckling Temperature of SSSS Square Plates ($\lambda_{cr} = 10^3 \alpha \Delta T$)

Source	$a/h = 10$	$a/h = 20$	$a/h = 100$
Zhang et al. (2014c)	11.99	3.123	0.1272
Wu et al. (2017a)	11.98	3.119	0.1265
Present	11.88	3.096	0.1255

Figure 8.1 shows the influence of the plate's aspect ratio on the variation of the critical buckling temperature of both UD and FG nanocomposite structures reinforced with GPLs. It can be seen from this diagram that the former estimations about the effect of the distribution patterns on the thermal response of the system are verified. In other words, because of the greater structural rigidity of the FG-X GPLR nanocomposite structures, such nanocomposite elements are able to endure higher thermal gradients. Therefore, such structures will buckle later than other types pf FG-GPLR nanocomposite plates. Among all types of nanocomposite plates, the weakest is the FG-O nanocomposite structure due to the fact that in this type, the concentration of reinforcements is in the layers close to the mid-surface of the

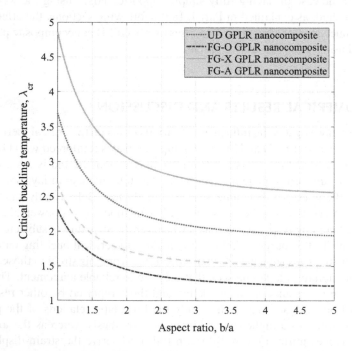

FIGURE 8.1 Variation of the critical buckling temperature of GPLR nanocomposite plates versus aspect ratio of the structure for various types of FG nanocomposites.

structure, which possess smaller cross-sectional rigidities. In addition, increasing the aspect ratio of the plate results in a more flexible structure which leads to a reduction in the buckling temperature of the plate. The reason for the decrease is the inverse relationship between flexibility (or compliance) and stiffness in linear elastic solids. Hence, it is natural to see that the buckling temperature of the nanocomposite plate will decrease in the case of using plates with high aspect ratios.

The influence of the ratio between the plate's length and its thickness on the thermal stability of the GPLR nanocomposite structures is shown in Figure 8.2. In this diagram, the variation of the critical buckling temperature against length-to-thickness ratio is plotted for various types of FG nanocomposites. Similar to the previous figure, the highest buckling temperature belongs to FG-X GPLR nanocomposite plates and the smallest one relates to those nanocomposite structures reinforced with O-pattern. Naturally, the buckling temperature decreases gradually as the length-to-thickness ratio of the plate increases. In this situation, the structural flexibility of the structure will increase as the plates move toward being a thin-walled structure. As mentioned before, the greater the flexibility of the plate, the greater will be its compliance and the stiffness of the structure will be lessened. So, it is not strange to see a decrease in the buckling response of the plate subjected to thermal gradient.

The coupled influence of the mode number and distribution pattern of the reinforcements on the variation of the critical buckling temperature of the GPLR

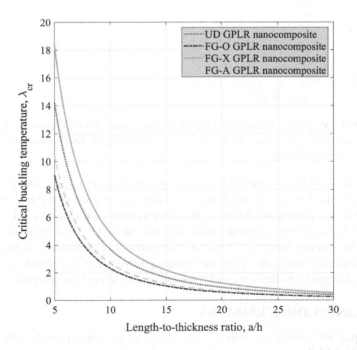

FIGURE 8.2   Variation of the critical buckling temperature of GPLR nanocomposite plates versus length-to-thickness ratio of the structure for various types of FG nanocomposites.

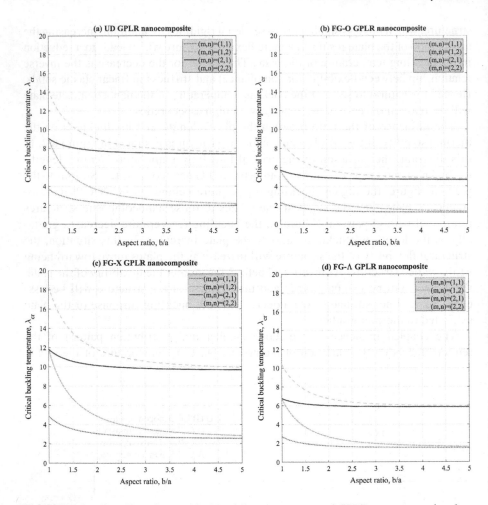

FIGURE 8.3 Variation of the critical buckling temperature of GPLR nanocomposite plate versus aspect ratio of the plate for various mode numbers and distribution patterns of the GPLs across the thickness.

nanocomposite plate is shown in Figure 8.3. It can be seen that increasing the aspect ratio results in a decrease in the critical buckling temperature of the nanocomposite plate due to the greater flexibility of the structure. Also, it can be easily seen that among all types of nanocomposite materials, the greatest critical buckling temperature can be attained when FG-X GPLR nanocomposite is utilized for the purpose of manufacturing the plate. Also, the buckling response of the structure in higher modes converges to its fundamental response as the aspect ratio increases.

8.3 CONCLUDING REMARKS

This chapter was concerned with the thermo-mechanical buckling analysis of multi-layered FG-GPLR nanocomposite plates regarding the influence of shear strain using the well-known Reddy-type plate hypothesis. The equivalent material properties of the nanocomposite plate were attained using the Halpin-Tsai micromechanical method. All of the edges of the plate were considered to be simply supported. The following points present the most significant highlights of the present study:

- The critical buckling temperature of the plate will lessen as the aspect ratio of the plate grows. This trend is due to the enhancement of the system's compliance when the aspect ratio of the plate grows.
- Similarly, increasing the length-to-thickness ratio of the nanocomposite plate will result in a reduction in the critical buckling temperature of the nanocomposite structure.
- Among various FG distribution patterns for the arrangement of the GPLs across the thickness of the plate, FG-X is the best one if the plate-type nanocomposite element will be subjected to high thermal gradients.

9 On Buckling Behaviors of FG-GOR Nanocomposite Plates Rested on Elastic Substrate

9.1 PROBLEM DEFINITION

In the present chapter, the elastic buckling problem of the rectangular FG-GOR nanocomposite plates will be solved in the framework of an analytic study. The effective material properties will be extracted using the Halpin-Tsai micromechanical procedure, explained in the Part 1 of the book. The plate-type element will be assumed to be embedded on a two-parameter elastic medium with Winkler and shear (Pasternak) springs. To present the problem in a mathematical format, the kinematic relations of the refined shear deformable plates will be incorporated with the concept of the principle of virtual work. The governing differential equations of the problem can be found referring to the Eqs. (2.215)–(2.218) once the terms including differentiation with respect to the time are ignored. Thereafter, the achieved governing equations of the problem will be solved on the basis of the well-known Navier-type solution to satisfy the simply supported BC at all of the edges of the plate.

9.2 NUMERICAL RESULTS AND DISCUSSION

Present part involves with the interpretation of the results of this chapter. The material properties of the matrix and nanosize reinforcing agents can be enriched using those reported in Chapter 3. To make it easier to understand the influence of each variant on the static stability responses of the FG-GOR nanocomposite plates, the following dimensionless form of the foundation's stiffnesses and buckling load is considered in all of the future illustrations:

$$K_w = k_w \frac{a^4}{D^*}, \quad K_p = k_p \frac{a^2}{D^*}, \quad N_{cr} = N^b \frac{a^2}{D^*}, \quad D^* = \frac{E_m h^3}{12\left(1 - v_m^2\right)} \quad (9.1)$$

The first numerical investigation can be found in Table 9.1, where the dimensionless buckling load of the FG-GOR nanocomposite plates is reported for different variants such as length-to-thickness ratio of the plate, weight fraction of the GOs,

TABLE 9.1

The Dimensionless Buckling Loads of FG-GOR Square Plates for Various Mode Numbers, Weight Fractions of GOs, and Length-To-Thickness Ratios ($K_w = 10$, $K_p = 1$)

W_{GO} (%)	a/h	(m, n)	UD GOR	FG-O GOR	FG-X GOR	FG-V GOR
0.2	10	(1, 1)	25.42766	22.98059	27.78171	24.98527
		(1, 2)	56.35875	51.32724	60.93252	55.45607
		(1, 3)	98.80538	91.39931	104.9293	97.45173
		(1, 4)	144.3481	135.9094	150.3783	142.6835
		(2, 2)	83.02943	76.34845	88.76943	81.72962
		(2, 3)	119.9685	111.8854	126.2653	118.3068
	20	(1, 1)	26.48173	23.77271	29.16425	25.99987
		(1, 2)	62.29402	55.85955	68.57361	61.18846
		(1, 3)	119.0487	107.2139	130.3258	117.0813
		(1, 4)	192.3015	174.354	208.8474	189.3468
		(2, 2)	96.81086	87.0226	106.2271	95.0413
		(2, 3)	151.2882	136.644	165.0453	148.6448
	50	(1, 1)	26.79398	24.00525	29.57823	26.30011
		(1, 2)	64.19603	57.28189	71.0828	63.02119
		(1, 3)	126.3451	112.7064	139.8767	124.1289
		(1, 4)	212.2309	189.487	234.6735	208.6338
		(2, 2)	101.5586	90.5873	112.4608	99.61077
		(2, 3)	163.3208	145.7364	180.7265	160.2411
0.5	10	(1, 1)	33.36404	27.16644	39.11838	31.28317
		(1, 2)	74.66279	61.66952	85.48574	70.40218
		(1, 3)	131.2406	111.499	144.9342	124.8233
		(1, 4)	191.9335	168.402	204.1445	184.0062
		(2, 2)	110.2155	92.63057	123.3542	104.0588
		(2, 3)	159.4438	137.4826	172.9951	151.5395
	20	(1, 1)	34.76661	27.97358	41.43287	32.50489
		(1, 2)	82.5611	66.34529	98.03758	77.36687
		(1, 3)	158.1824	128.1102	185.5992	148.9254
		(1, 4)	255.763	209.6384	295.2352	241.8351
		(2, 2)	128.5564	103.766	151.5737	120.2329
		(2, 3)	201.1296	163.7368	234.3056	188.6766
	50	(1, 1)	35.18209	28.2089	42.13367	32.86546
		(1, 2)	85.0919	67.78921	102.2634	79.58028
		(1, 3)	167.8912	133.7144	201.5558	157.4918
		(1, 4)	282.283	225.1825	337.9433	265.3979
		(2, 2)	134.8738	107.3959	162.021	125.7338
		(2, 3)	217.1411	173.0411	260.3857	202.6858
1	10	(1, 1)	46.58117	34.06309	57.8191	40.6804
		(1, 2)	105.1512	78.5046	125.5746	93.01043
		(1, 3)	185.2792	143.7831	209.5579	166.8781
		(1, 4)	271.2324	220.052	290.345	248.3515

TABLE 9.1 *(Continued)*
The Dimensionless Buckling Loads of FG-GOR Square Plates for Various Mode Numbers, Weight Fractions of GOs, and Length-To-Thickness Ratios ($K_w = 10$, $K_p = 1$)

W_{GO} (%)	a/h	(m, n)	UD GOR	FG-O GOR	FG-X GOR	FG-V GOR
		(2, 2)	155.505	118.902	179.4204	137.8723
		(2, 3)	225.2189	178.3784	248.1818	202.4435
	20	(1, 1)	48.5628	34.9487	61.80874	42.16828
		(1, 2)	116.312	83.6858	146.7979	101.6066
		(1, 3)	223.357	162.4602	276.596	197.0956
		(1, 4)	361.4645	267.2275	436.5972	321.848
		(2, 2)	181.4249	131.3526	226.3766	157.8218
		(2, 3)	284.1409	208.1292	348.0151	248.7502
	50	(1, 1)	49.14977	35.20551	63.03075	42.60572
		(1, 2)	119.8875	85.26545	154.1274	104.3156
		(1, 3)	237.0744	168.6157	304.0443	207.6833
		(1, 4)	398.937	284.3922	509.3031	351.193
		(2, 2)	190.3504	135.3333	244.4058	164.5186
		(2, 3)	306.7639	218.3727	392.6716	265.8947

and mode numbers. Clearly, the greatest stability endurance corresponds to the FG-X GOR nanocomposite plates as we knew before. This type of reinforcing pattern increases the cross-sectional rigidities of the plate-type element due to high values at the top and bottom edges. It is evident that increasing the length-to-thickness ratio results in an increase in the dimensionless buckling response of the plate. This trend seems ridiculous at first; however, it is completely correct. In fact, the results reported in Table 9.1 are in the dimensionless form and by paying attention to the style of the dimensionless form of the buckling load it can be found that the square of the plate's length is multiplied in the buckling load. So, it can be proven that increasing the plate's length results in an increase in the dimensionless buckling load of the plate.

Figure 9.1 plots the variation of the dimensionless buckling load of the GOR nanocomposite plates versus the Winkler coefficient for various types of FG nanocomposite materials. It is clear that increasing the stiffness of the Winkler coefficient will be result in an increase in the dimensionless response of the plate because the equivalent stiffness of the plate will be increased by using stiffer springs in the elastic foundation of the structure. According to this diagram, the FG-X GOR nanocomposite structures can endure greater excitations in comparison with the other types of polymeric nanocomposite structures. Obviously, the once the GOs are added to the resin using either FG-O or FG-V pattern, the stiffness of the structure will be reduced consequently.

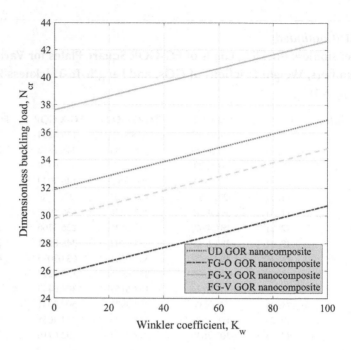

FIGURE 9.1 Variation of the dimensionless buckling load of the GOR nanocomposite square plates versus Winkler coefficient for various types of FG nanocomposites ($a/h = 10$, $W_{GO} = 0.5\%$, $K_p = 0$).

A similar illustration is presented in Figure 9.2, drawing the variation of the dimensionless buckling load of the GOR nanocomposite plates against the Pasternak coefficient for various types of FG-GOR nanocomposites. The highlights of this figure are as same as those of the previous one and the only difference is in the magnitude of the variations. In other words, the dimensionless buckling load of the plate can be increased once the magnitude of the Pasternak coefficient is raised. It is interesting to point out that the slope of the buckling curve of the plate is higher in the case of tuning the Pasternak coefficient compared to the situation where the magnitude of the Winkler term is varied. Therefore, the impact of the Pasternak coefficient is many times greater than that of the Winkler coefficient. Another observation of this figure concerns the distribution patterns of FG-GOR nanocomposite materials, which were expressed in the interpretation of the former illustration.

The main purpose of presenting Figure 9.3 is to show how the dimensionless buckling response of the GOR nanocomposite plates can be enlarged by changing the weight fraction of the reinforcing phase. Based on this diagram, it can be seen that increasing the weight fraction of the GOs can be one of the best ways to strengthen the plate-type element. Indeed, adding a tiny amount to the content of the GOs results in a remarkable increase in the buckling endurance of the plate. This marvelous reinforcing capacity comes from the enhanced material properties of carbonic agents in the nanoscale. Indeed, the improvement of the composites' stiffness can be performed in the best form using nanosize GOs. Similar to the previous

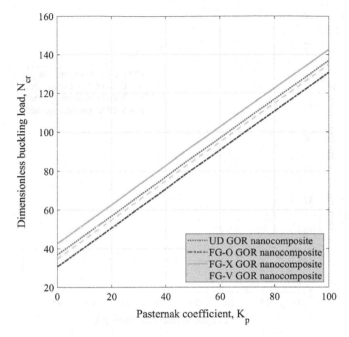

FIGURE 9.2 Variation of the dimensionless buckling load of the GOR nanocomposite square plates versus Pasternak coefficient for various types of FG nanocomposites ($a/h = 10$, $W_{GO} = 0.5\%$, $K_w = 100$).

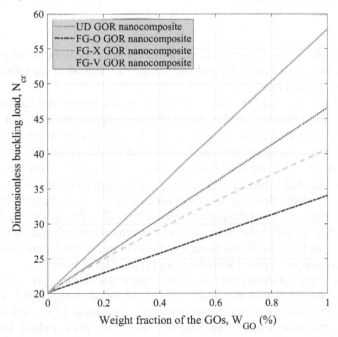

FIGURE 9.3 Variation of the dimensionless buckling load of the GOR nanocomposite square plates versus weight fraction of the GOs for various types of FG nanocomposites ($a/h = 10$, $K_w = 10$, $K_p = 1$).

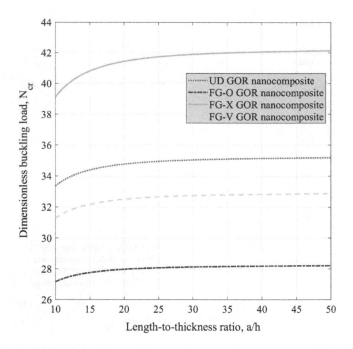

FIGURE 9.4 Variation of the dimensionless buckling load of the GOR nanocomposite square plates versus length-to-thickness ratio of the plate for various types of FG nanocomposites ($W_{GO} = 0.5\%$, $K_w = 10$, $K_p = 1$).

illustrations, this figure shows once again that the stability limit of the plate peaks when the X-type distribution is employed for the distribution of the GOs across the thickness of the structure.

Figure 9.4 investigates the influence of the plate's length-to-thickness ratio on the dimensionless buckling load of the GOR nanocomposite structures once various types of distribution patterns are utilized for the goal of reinforcing the polymeric matrix. Based on this figure, the stability load that can be tolerated by the plate will be increased once the length-to-thickness ratio of the plate becomes greater. This trend is a result of the direct relationship between the dimensionless buckling load and the plate's length. However, this increasing influence can be better seen when the length-to-thickness ratio of the plate is smaller than 30. After that point, the dimensionless buckling load of the plate remains constant and the changes will be negligible. Again, the FG-X GOR nanocomposite structures possess bigger dimensionless buckling responses compared with the other types of GOR plates.

Next, the impact of the aspect ratio of the plate under observation on the buckling responses of the GOR nanocomposite plate is displayed in Figure 9.5. This figure reveals that increasing the ratio between the plate's width and its length results in a reduction in the dimensionless buckling response of the plate. Indeed, the effect of the width of the plate will have a higher impact than that of its length. Hence, it is natural to see a decrease in the buckling load of the plate due to its direct relation

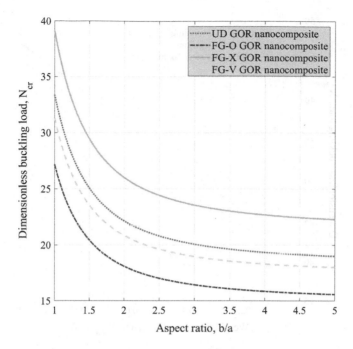

FIGURE 9.5 Variation of the dimensionless buckling load of the GOR nanocomposite plates versus aspect ratio of the plate for various types of FG nanocomposites ($a/h = 10$, $W_{GO} = 0.5\%$, $K_w = 10$, $K_p = 1$).

with length. Similar to previous figure, the effect of the aspect ratio will dissipate whenever this ratio becomes greater than four. Indeed, the most crucial dropdown occurs when this ratio is changed from one to two. After this change, the amplitude of the variation of the dimensionless buckling load in further changes is small.

As the final example, the variation of the dimensionless buckling load of the UD GOR nanocomposite plates against the weight fraction of the GOs is plotted in Figure 9.6 for various types of elastic seats for the plate-type element. This figure summarizes the former illustrations. It is clear that increasing the weight fraction of the GOs results in an increase in the dimensionless buckling load of the nanocomposite plate. In addition, the effect of changing the stiffness of the springs of the elastic substrate on the static stability responses of the plate can be monitored in this diagram. In other words, increasing the stiffness of each of the Winkler or Pasternak coefficients results in an enhancement in the total stiffness of the plate which results in the plate tolerating higher static forces.

9.3 CONCLUDING REMARKS

This chapter was concerned with solving the static stability problem of polymeric nanocomposite plates reinforced with an FG distribution of GOs across the thickness direction. In this analysis, the effective material properties were obtained according to a micromechanical method, called the Halpin-Tsai method. On the other hand,

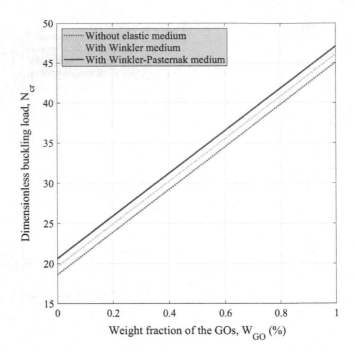

FIGURE 9.6 Variation of the dimensionless buckling load of the UD GOR nanocomposite square plates versus weight fraction of the GOs for various types of elastic medium used as the structure's seat ($a/h = 10$).

the kinematic relations of the plate were derived with a refined-type higher-order shear deformation plate hypothesis to be able to analyze plates with small length-to-thickness ratios regarding the effects of shear deflection. Assuming that the plate is simply supported at all of the edges, the governing equations were solved employing a Navier-type analytical method. The following points present the most significant highlights of this chapter:

- The dimensionless buckling load of the nanocomposite plate is increased when the FG-X pattern is implemented for the arrangement of the GOs across the thickness of the plate. On the other hand, FG-O GOR nanocomposite plates demonstrate the minimum stability endurance.
- The higher the content of the nanosize GO reinforcements, the higher will be the stability load of the GOR nanocomposite structure.
- Increasing each of the Winkler or Pasternak springs results in an improvement in the buckling limit of the GOR nanocomposite plates.

10 Thermal Buckling Analysis of Embedded GOR Nanocomposite Plates

10.1 PROBLEM DEFINITION

This chapter investigates the thermo-elastic buckling problem of FG-GOR nano-composite plate-type elements once the plate is assumed to be subjected to different types of temperature rise such as uniform, linear, and sinusoidal ones. The thermo-mechanical material properties of the single-layered plate are developed using the Halpin-Tsai micromechanical homogenization procedure. In order to formulate the problem, the displacement field of the refined shear deformable plate theories is combined with the definition of the linear strain tensor in the continuum mechanics to determine the strain-displacement relationships of the plate regarding for the influence of the shear strains. Afterward, the governing equations of the problem will be developed on the basis of the principle of virtual work for the static stability problem. The set of the governing PDEs of this problem can be found by referring to Eqs. (2.215)–(2.218), dismissing the time-dependent terms due to the static nature of the proposed analysis. In this chapter, the critical buckling temperature results will be presented instead of the critical buckling load. In other words, the buckling mode in the continuum is presumed to be excited due to the existence of thermal gradient. It will be shown that the effects of various distribution patterns of reinforcing of the plate on the buckling temperature of the structure can be deeply changed when the type of the applied temperature rise is varied.

10.2 NUMERICAL RESULTS AND DISCUSSION

This section includes five illustrative numerical examples showing the influence of various terms on the thermal stability responses of FG-GOR nanocomposite plates. The primary inputs of the analysis (i.e., the thermo-mechanical material properties of the polymeric matrix and reinforcing phase) can be considered to be as same as those reported in Chapter 4. Before discussing the phenomenological trends that are achieved throughout this chapter, it is better to introduce the following dimensionless form of the parameters for the sake of simplicity:

$$K_w = k_w \frac{a^4}{D^*}, \quad K_p = k_p \frac{a^2}{D^*}, \quad \lambda_{cr} = 1000\alpha_m\Delta T, \quad D^* = \frac{E_m h^3}{12\left(1 - v_m^2\right)} \quad (10.1)$$

where λ_{cr} is the critical buckling temperature of the plate; also, K_w and K_p are the dimensionless forms of the Winkler and Pasternak coefficients of the elastic medium, respectively.

The variation of the critical buckling temperature of the GOR nanocomposite plate versus Winkler coefficient is drawn in Figure 10.1 for various types of FG distributions for the reinforcing GOs. It is evident that increasing the coefficient of the Winkler springs of the elastic substrate leads to an increase in the dimensionless buckling temperature of the plate. Indeed, the structural stiffness of the plate will be enhanced, and, due to this reality, the structure will be strengthened and it will be able to tolerate higher buckling temperatures. One other crucial highlight is the dependency of the critical buckling temperature of the plate and the type of the temperature rise once FG nanocomposites are employed. In fact, the previously

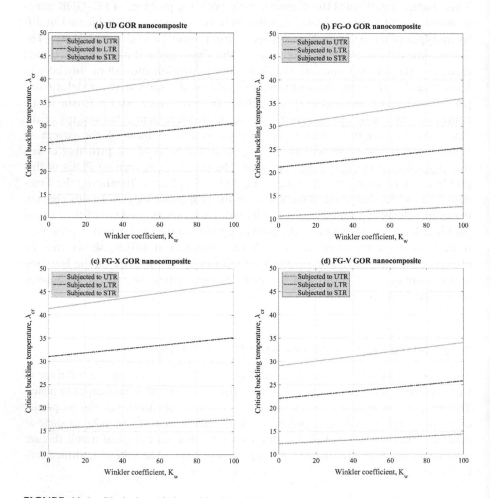

FIGURE 10.1 Variation of the critical buckling temperature of square GOR nanocomposite plates versus Winkler coefficient for different types of temperature rise ($a/h = 10$, $W_{GO} = 0.5\%$, $K_p = 0$).

known order for the mechanical response of the system which can be observed in the UTR nanocomposite plates (i.e., observing the peak response for the FG-X GOR nanocomposites followed by UD GOR, FG-V GOR, and FG-O GOR ones) may be not found in the FG-GOR nanocomposite plates. In other words, once either LTR or STR is applied to the structure, the FG-O GOR nanocomposites can reveal higher thermal stability endurance than the FG-V GOR ones. Also, plates manufactured from FG-X GOR nanocomposites and UD GOR ones are the best alternatives once any arbitrary temperature rise is applied. Furthermore, among all types of nanocomposite structures, the worst thermal stability condition can be reported in the case of applying UTR on the plate followed by LTR and STR, respectively.

The variation of the critical buckling temperature versus Pasternak coefficient for different types of temperature rise can be seen in Figure 10.2. The trends observed in

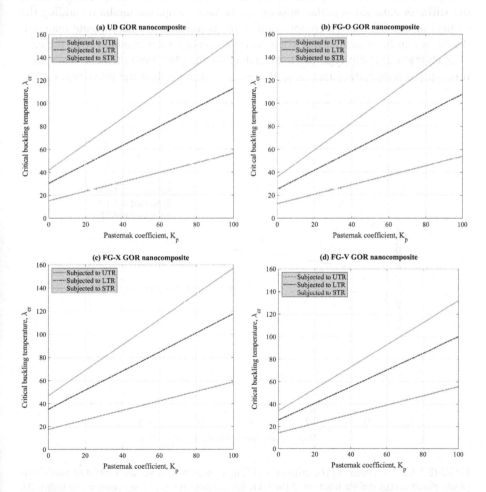

FIGURE 10.2 Variation of the critical buckling temperature of square GOR nanocomposite plates versus Pasternak coefficient for different types of temperature rise ($a/h = 10$, $W_{GO} = 0.5\%$, $K_w = 100$).

the previous illustration can be seen in this diagram, too. Indeed, the thermal buckling responses of the nanocomposite plate can be improved using stiffer Pasternak springs in the seat of the structure. Comparing the thermal buckling curves of this figure and the previous one, it can be concluded that the Pasternak coefficient can amplify the thermo-elastic response of the system many times more than the Winkler coefficient. It is worth mentioning that the dependency of the critical buckling temperature and the type of temperature rise for various types of FG nanocomposites can be seen in this figure, too.

In the next example, the variation of the critical buckling temperature of UD GOR nanocomposite square plates versus weight fraction of the GOs is drawn (see Figure 10.3). Based on this diagram, it can be determined that an increase in the weight fraction of the GOs can result in an increase in the critical buckling temperature of the GOR nanocomposite structure. The reason for this amplifying trend is the stiffness enhancement that appears in the nanocomposite media by adding the content of the GOs in the composition of the constituent nanocomposite material. Again, it can be seen that the worst thermal condition corresponds with the case of applying a uniform-type thermal loading on the plate-type element. Therefore, this type of temperature rise must be recognized more than the other types when

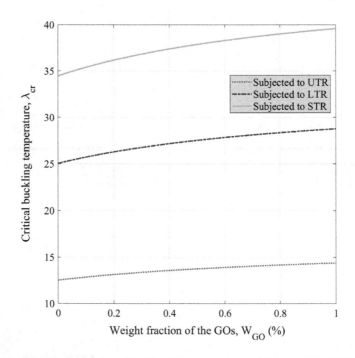

FIGURE 10.3 Variation of the critical buckling temperature of square UD GOR nanocomposite plates versus weight fraction of the GOs for different types of temperature rise ($a/h = 10$, $K_w = 10$, $K_p = 1$).

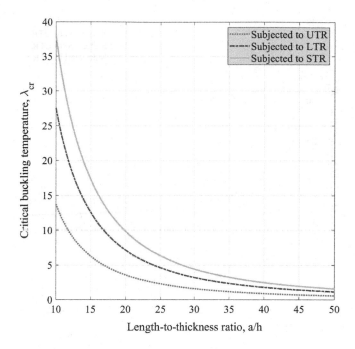

FIGURE 10.4 Variation of the critical buckling temperature of square UD GOR nanocomposite plates versus length-to-thickness ratio of the plate for different types of temperature rise ($W_{GO} = 0.5\%$, $K_w = 10$, $K_p = 1$).

designing a GOR nanocomposite plate-type device that is going to be able to tolerate enormous thermal gradients.

Figure 10.4 probes the influence of the length-to-thickness ratio of the GOR nanocomposite plates on the thermal buckling responses of UD nanocomposite structures. On the basis of this figure, it can be concluded that increasing the length-to-thickness ratio of the plate will be resulted in a reduction in the critical buckling temperature of the GOR nanocomposite system. This can be justified by noting the effect of increasing the length-to-thickness ratio of the plate on its mechanical behavior. In fact, the structure will be more flexible once its length-to-thickness ratio is added. Recall that the greater the flexibility of the system, the more its compliance will be increased. Also, the inverse relationship between compliance and stiffness can be observed in linear elastic solids. Therefore, due to the reverse relation between the compliance and stiffness, it can be concluded that the stiffness of the plate will be reduced in the case of using high length-to-thickness ratios. It is worth mentioning that the effect of the length-to-thickness ratio on the critical buckling temperature will be dissipated as this ratio exceeds 40. In other words, after this particular ratio, changing the length-to-thickness ratio does not possess a remarkable impact on the buckling temperature of the plate. As we knew

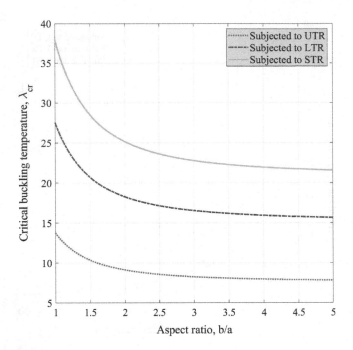

FIGURE 10.5 Variation of the critical buckling temperature of square UD GOR nanocomposite plates versus aspect ratio of the plate for different types of temperature rise ($a/h = 10$, $W_{GO} = 0.5\%$, $K_w = 10$, $K_p = 1$).

before, the greatest buckling temperature belongs to the plates subjected to STR followed by LTR and UTR, respectively.

As the final illustration of this chapter, the variation of the critical buckling temperature of the UD GOR nanocomposite plates is plotted versus aspect ratio of the plate in Figure 10.5. The figure reveals that increasing the aspect ratio of the structure can result in a decrease in the critical buckling temperature of the continuous system. The reason is the reduction of the structural stiffness of the nanocomposite plate due to the increase in the system's flexibility once the aspect ratio is increased. Also, it is clear that the plates subjected to UTR can endure lower thermal gradients compared with ones subjected to LTR and STR. Similar to the previous figure, it is shown that the effect of increasing the aspect ratio of the plate on the thermal buckling response of the system cannot be observed in aspect ratios greater than 4.

10.3 CONCLUDING REMARKS

The main concern of this chapter was to analyze the thermal buckling problem of GOR single-layered nanocomposite plates resting on a two-parameter elastic foundation when they are subjected to various temperature profiles. Extending the Halpin-Tsai micromechanical method for GOR nanocomposites, the thermo-mechanical material properties of the nanocomposites under observation were attained. The combination of the principle of virtual work and the refined plate theories resulted in deriving the governing equations of the problem in terms of the displacement field of the plate. The most significant highlights of this chapter follow:

- The critical buckling temperature of the FG-GOR nanocomposite structures can be enhanced using a larger amount of the GOs in the nanocomposite's composition.
- The critical buckling temperature can be intensified by increasing each of the Winkler or Pasternak coefficients.
- Among all types of temperature rise, the worst thermal stability situation will happen when the nanocomposite plate is subjected to UTR followed by LTR and STR, respectively.
- Except for UTR, the critical buckling temperature of the FG-O GOR nanocomposite plates is larger than that of the FG-V GOR ones.
- The thermal stability response of the GOR nanocomposite plate can be reduced by increasing either the length-to-thickness ratio or the aspect ratio of the plate.

Part 3

Dynamic Analyses

Part 3

Dynamic Analyses

11 Thermo-Mechanical Vibration Analysis of Sandwich Beams with CNTR Nanocomposite Facesheets

11.1 PROBLEM DEFINITION

In this chapter, the thermo-mechanical vibrational responses of sandwich beams will be for a beam consisting of an elastic core placed between two CNTR nanocomposite facesheets. The equivalent material properties of the nanocomposite facesheets will be developed using the modified form of the rule of the mixture. Via this method, the influences of the tiny dimensions of the CNTs will be covered, introducing size-dependent modifying coefficients. The kinematic relations of the beam will be derived based on the well-known Timoshenko beam hypothesis. Thereafter, the dynamic form of the principle of virtual work will be implemented to derive the governing equations of the beam for the case of thermal vibration problem. The set of the governing equations can be found by referring to Eqs. (2.112)–(2.114). The natural frequency of the system will be determined by employing the DTM considering various types of edge supports. Since FG materials like FG-CNTR nanocomposites are generally used in the cases of severe thermal gradients, the material properties of the core and the nanocomposite facesheets are assumed to be as functions of the local temperature to reach more reliable thermal results. In this chapter, the dependency of the material properties of the CNTs and the core of the sandwich structure on the temperature will be tailored using the following nonlinear equation (Shen, 2004):

$$P = P_0 \left(P_{-1}T^{-1} + 1 + P_1T + P_2T^2 + P_3T^3 \right) \tag{11.1}$$

in which P_is are the temperature-dependent coefficients which are presented in Table 11.1 for the homogeneous material which is used in the core of the sandwich beam. Also, the temperature-dependent coefficients of the nanosize CNTs are presented in Table 11.2. It is worth mentioning that poly(methyl methacrylate) (PMMA) is utilized as the host of the CNTR nanocomposite and the temperature-dependency of its material properties is considered in

TABLE 11.1

Temperature-Dependent Properties of Young's Modulus and Thermal Expansion Coefficient for the Homogeneous Core

Material	Properties	P_0	P_{-1}	P_1	P_2	P_3
Ti-6Al-4V	$E(Pa)$	122.56e+9	0	-4.586e-4	0	0
	$\alpha(K^{-1})$	7.5788e-6	0	6.638e-4	-3.147e-6	0

this chapter using the following expressions for its Young's moduli and CTE (Yang et al., 2015):

$$E_m = 3.52 - 0.0034T \ (\text{GPa}),$$
$$\alpha_m = 45(1 + 0.0005\Delta T) \ (1/\text{K})$$

11.2 NUMERICAL RESULTS AND DISCUSSION

In this section a group of numerical tables and figures will be presented to show the influence of various terms on the thermo-elastic fluctuation responses of sandwich beams stiffened via nanocomposite facesheets. The upper and lower facesheets are constructed by dispersion of armchair (10, 10) CNTs in a polymeric matrix. In this investigation, the PMMA is presumed to be the initial matrix which has the following material properties:

$$E_m = 2.5 \ \text{GPa}, \quad \rho_m = 1190 \ \text{kg}/\text{m}^3, \quad v_m = 0.3$$

As stated in the previous paragraphs, the efficiency parameters, utilized in the homogenization procedure, must be obtained comparing the mechanical properties of the nanocomposite achieved from the rule of mixture with those enriched from

TABLE 11.2

Temperature-Dependent Properties of Young's Modulus and Thermal Expansion Coefficient for CNTs

Temperature (K)	E_{11}^{CNT} (TPa)	E_{22}^{CNT} (TPa)	G_{12}^{CNT} (TPa)	α^{CNT} (1 / K)
300	5.6466	7.0800	1.9445	3.4584
500	5.5308	6.9348	1.9643	4.5361
700	5.4744	6.8641	1.9644	4.6677

the MD (Han and Elliott, 2007). According to the above equations, the efficiency parameters can be assumed to be as (Shen and Zhang, 2010):

$$\eta_1 = 0.137, \eta_2 = 1.022, \eta_3 = 0.715 \quad \text{for } V_{CNT}^* = 0.12$$
$$\eta_1 = 0.142, \eta_2 = 1.626, \eta_3 = 1.138 \quad \text{for } V_{CNT}^* = 0.17$$
$$\eta_1 = 0.141, \eta_2 = 1.585, \eta_3 = 1.109 \quad \text{for } V_{CNT}^* = 0.28$$

Also, titanium alloy with moduli of $E_c = 113.8$ GPa, density of $\rho_c = 4430 \text{ kg} / \text{m}^3$, and Poisson's ratio of $v_c = 0.342$ is chosen for the core of the sandwich beam (Ti-6Al-4V). The entire thickness of the sandwich beam is assumed to be 10 mm and will not change throughout the examples. The ratio between the core's thickness and that of the face sheets will vary over various numerical examples. This ratio will be considered to be $h_c / h_f = 4, 6, 8$ during the numerical analyses.

In this section we will show the efficiency of the presented semi-analytical solution method in predicting the natural frequency values of beam-type elements. Table 11.3

TABLE 11.3

Convergence Study for the First Three Frequencies with FG-CNTRC Facesheets ($L/h = 20$, $h_c/h_f = 8$)

n	ω_1	ω_2	ω_3
12	0.14499	-	-
13	0.14502	-	-
14	0.14503	-	-
15	0.14504	0.54167	-
16	0.14504	0.59675	0.72903
17	0.14504	0.58092	0.83827
18	0.14504	0.57080	42.7243
19	0.14504	0.57184	5.0476
20	0.14504	0.57289	1.05913
21	0.14504	0.57279	1.11939
22	0.14504	0.57269	41.59312
23	0.14504	0.57270	6.07850
24	0.14504	0.57270	1.23659
25	0.14504	0.57270	1.24982
26	0.14504	0.57270	1.26681
27	0.14504	0.57270	1.26417
28	0.14504	0.57270	1.26180
29	0.14504	0.57270	1.26206
30	0.14504	0.57270	1.26232
31	0.14504	0.57270	1.26227
32	0.14504	0.57270	1.26227
33	0.14504	0.57270	1.26227

TABLE 11.4

Comparison of the First Three Dimensionless Natural Frequencies of S-S Sandwich Beams with FG-CNTR Nanocomposite Facesheets ($L/h = 20$, $h_c/h_f = 8$)

Mode		$V_{CNT}^* = 0.12$		$V_{CNT}^* = 0.17$		$V_{CNT}^* = 0.28$	
		Present	(Wu et al., 2015)	Present	(Wu et al., 2015)	Present	(Wu et al., 2015)
1	FG	0.1450	0.1453	0.1594	0.1588	0.1844	0.1825
	UD	0.1429	0.1432	0.1566	0.1560	0.1806	0.1785
2	FG	0.5727	`0.5730	0.6289	0.6247	0.7261	0.7174
	UD	0.5643	0.5650	0.6180	0.6140	0.7114	0.6997
3	FG	1.2623	1.2599	1.3837	1.3689	1.5933	1.5554
	UD	1.2444	1.2429	1.3605	1.3465	1.5623	1.5246

shows the number of required iterations to reach the converged response for the first three natural modes of the beam's oscillation. It can be inferred that the converged natural frequency of the 1st, 2nd, and 3rd modes can be obtained by the means of using 15, 23, and 31 iterations in the implemented DTM. The next verification studies are displayed in Table 11.4 and Table 11.5 to compare the natural frequencies of a sandwich beam with those reported in the open literature by (Wu et al., 2015) for two types of edge supports, namely S-S and C-C. It can be observed that the frequency values resulting from our semi-analytical method are compatible with those developed by the researchers in a previously published article. Hence, it can be inferred that the employed DTM is powerful enough to estimate the frequency of the beam-type elements with a remarkable precision.

TABLE 11.5

First Three Dimensionless Natural Frequencies of C-C Sandwich Beams with FG-CNTRC Facesheets ($L/h = 20$, $h_c/h_f = 8$)

Mode		$V_{CNT}^* = 0.12$		$V_{CNT}^* = 0.17$		$V_{CNT}^* = 0.28$	
		Present	(Wu et al., 2015)	Present	(Wu et al., 2015)	Present	(Wu et al., 2015)
1	FG	0.3239	0.3240	0.3528	0.3530	0.4031	0.4032
	UD	0.3192	0.3195	0.3467	0.3470	0.3950	0.3949
2	FG	0.8724	0.8704	0.9483	0.9443	1.0800	1.0699
	UD	0.8602	0.8588	0.9327	0.9291	1.0594	1.0492
3	FG	1.6626	1.6520	1.8026	1.7838	2.0441	2.0029
	UD	1.6404	1.6313	1.7744	1.7569	2.0086	1.9672

TABLE 11.6

Effect of Nanotube Volume Fraction on the First Three Natural Frequencies of Sandwich Beams with FG-CNTR Nanocomposite Facesheets (L/h = 20, h_c/h_f = 8)

Mode	BC		$\Delta T = 0$ V_{cn}^*			$\Delta T = 200$ V_{cn}^*			$\Delta T = 400$ V_{cn}^*		
			0.12	0.17	0.28	0.12	0.17	0.28	0.12	0.17	0.28
1	S-S	FG	0.1450	0.1595	0.1844	0.1393	0.1538	0.1789	0.1340	0.1487	0.1741
	S-S	UD	0.1429	0.1566	0.1806	0.1370	0.1509	0.1749	0.1317	0.1457	0.1699
2	S-S	FG	0.5727	0.6289	0.7261	0.5518	0.6086	0.7065	0.5319	0.5899	0.6893
	S-S	UD	0.5643	0.6180	0.7114	0.5432	0.5976	0.6917	0.5231	0.5785	0.6741
3	S-S	FG	1.2623	1.3837	1.5933	1.2166	1.3394	1.5505	1.1725	1.2980	1.5124
	S-S	UD	1.2444	1.3605	1.5623	1.1983	1.3159	1.5193	1.1539	1.2740	1.4807
1	C-C	FG	0.3239	0.3528	0.4031	0.3121	0.3414	0.3922	0.3008	0.3309	0.3826
	C-C	UD	0.3192	0.3467	0.3950	0.3072	0.3353	0.3841	0.2958	0.3245	0.3742
2	C-C	FG	0.8724	0.9483	1.0800	0.8407	0.9177	1.0508	0.8100	0.8891	1.0246
	C-C	UD	0.8602	0.9327	1.0594	0.8283	0.9019	1.0300	0.7973	0.8729	1.0034
3	C-C	FG	1.6626	1.8026	2.0441	1.6003	1.7425	1.9891	1.5423	1.6881	1.9379
	C-C	UD	1.6404	1.7744	2.0086	1.5778	1.7140	1.9505	1.5193	1.6579	1.8989
1	C-S	FG	0.2251	0.2474	0.2857	0.2167	0.2391	0.2778	0.2088	0.2316	0.2708
	C-S	UD	0.2218	0.2430	0.2799	0.2133	0.2347	0.2718	0.2052	0.2270	0.2647
2	C-S	FG	0.7166	0.7860	0.9059	0.6905	0.7607	0.8814	0.6655	0.7372	0.8597
	C-S	UD	0.7063	0.7727	0.8880	0.6801	0.7472	0.8634	0.6548	0.7234	0.8414
3	C-S	FG	1.4584	1.5962	1.8337	1.4052	1.5446	1.7838	1.3538	1.4961	1.7389
	C-S	UD	1.4383	1.5703	1.7992	1.3848	1.5184	1.7492	1.3329	1.4694	1.7039

In Table 11.6 and Figure 11.1, the combined impacts of temperature rise, mode number, BC, and volume fraction of the CNTs on the vibrational responses of sandwich beams stiffened via both FG and uniform distributions of nanofillers across the thickness are displayed. It is clear from both tabulated and graphical examples that the greatest natural frequencies can be developed whenever the C-C type of the investigated BCs is selected. The reason for this is that the structural stiffness of the system will be enlarged in this case followed by C-S and S-S types of BCs. On the other hand, it is clear that the natural frequency of the sandwich beam will decrease once the element is placed in an environment with nonzero thermal gradient. In fact, the higher the temperature rise, the softer will be the beam and due to this reciprocal influence, the natural frequency will be reduced for higher values of temperature rise. Among all of the studied mode numbers, the greatest natural frequency can be achieved when the volume fraction of the CNTs grows. The main reason is the superior stiffness of the CNT in comparison with that of the matrix that is employed to be the host of the nanofillers. In other words, from the greater moduli of the CNT in comparison with the matrix it can be concluded that the entire stiffness of the face sheets can be dramatically intensified by increasing the volume fraction

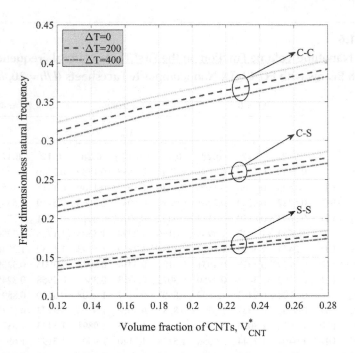

FIGURE 11.1 First three dimensionless natural frequencies of C-C, S-S, and C-S sandwich beams with FG-CNTR nanocomposite facesheets with different CNT volume fraction.

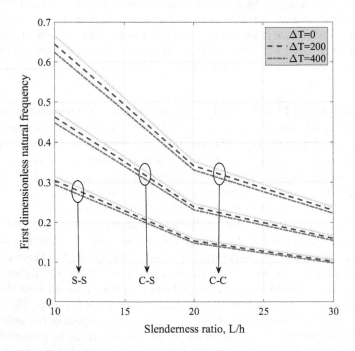

FIGURE 11.2 First natural frequency of C-C, S-S, and C-S sandwich beams with FG-CNTR nanocomposite facesheets with different slenderness ratios.

TABLE 11.7

Dimensionless First Three Natural Frequencies of Sandwich Beams with FG-CNTR Nanocomposite Facesheets and Different Values of Slenderness Ratio ($h_c/h_f = 8$, $V_{CNT}^* = 0.17$)

Mode	BC		$\Delta T = 0$ L/h			$\Delta T = 200$ L/h			$\Delta T = 400$ L/h		
			10	20	30	10	20	30	10	20	30
1	S-S	FG	0.3145	0.1595	0.1066	0.3043	0.1538	0.1021	0.2945	0.1487	0.0983
	S-S	UD	0.3090	0.1566	0.1047	0.2988	0.1509	0.1001	0.2893	0.1457	0.0962
2	S-S	FG	1.1943	0.6289	0.4237	1.1556	0.6086	0.4094	1.1192	0.5899	0.3965
	S-S	UD	1.1752	0.6180	0.4162	1.1363	0.5976	0.4018	1.0995	0.5785	0.3886
3	S-S	FG	2.4953	1.3837	0.9434	2.4116	1.3394	0.9130	2.3315	1.2980	0.8849
	S-S	UD	2.4597	1.3605	0.9270	2.3757	1.3159	0.8964	2.2952	1.2740	0.8678
1	C-C	FG	0.6661	0.3528	0.2379	0.6443	0.3414	0.2300	0.6237	0.3309	0.2227
	C-C	UD	0.6557	0.3467	0.2337	0.6339	0.3353	0.2257	0.6131	0.3245	0.2183
2	C-C	FG	1.6884	0.9483	0.6483	1.6308	0.9177	0.6272	1.5753	0.8891	0.6078
	C-C	UD	1.6657	0.9327	0.6371	1.6078	0.9019	0.6159	1.5521	0.8729	0.5962
3	C-C	FG	3.0220	1.8026	1.2518	2.9148	1.7425	1.2115	2.8114	1.6881	1.1747
	C-C	UD	2.9874	1.7744	1.2308	2.8794	1.7140	1.1902	2.7758	1.6579	1.1529
1	C-S	FG	0.4781	0.2474	0.1660	0.4626	0.2391	0.1599	0.4482	0.2316	0.1547
	C-S	UD	0.4702	0.2430	0.1630	0.4547	0.2347	0.1596	0.4400	0.2270	0.1515
2	C-S	FG	1.4468	0.7860	0.5331	1.3987	0.7607	0.5156	1.3529	0.7372	0.4995
	C-S	UD	1.4254	0.7727	0.5238	1.3772	0.7472	0.5061	1.3311	0.7234	0.4898
3	C-S	FG	2.7758	1.5962	1.0983	2.6799	1.5446	1.0629	2.5874	1.4961	1.0301
	C-S	UD	2.7397	1.5703	1.0795	2.6438	1.5184	1.0439	2.5510	1.4694	1.0107

of the CNTs. Regarding the direct relation between the system's stiffness and the natural frequency, it can be found that the natural frequency will be amplified as the volume fraction of the nanofillers increases.

In addition, the effects of slenderness ratio, temperature rise, mode number, distribution pattern of the CNTs, and BCs on the natural frequency behaviors of sandwich beams are included in Table 11.7 and Figure 11.2. It is clear that the previous findings can be observed again. Indeed, increasing the temperature rise will result in a reduction in the frequency. In addition, the FG distribution of the nanosize reinforcements of the face sheet causes an increase in the natural frequency compared with the case of implementing UD. The reason of this trend is that by using the FG pattern, the axes which have greatest moment radii's will be strengthened and this will result in an increase in the cross-sectional rigidities, which leads to stiffness enhancement. Again, the natural frequency becomes greater in the case of utilization of the C-C BC followed by C-S and S-S types. The most significant trend that can be observed in this table and figure is the reducing path the natural frequency goes through while the slenderness ratio becomes greater. The physical reason is that the structure will be

TABLE 11.8

Dimensionless First Three Natural Frequencies of Sandwich Beams with FG-CNTR Nanocomposite Facesheets and Various Values of Core-to-Facesheet Thickness Ratio ($L/h = 20$, $V^*_{CNT} = 0.17$)

Mode	BC		$\Delta T = 0$			$\Delta T = 200$			$\Delta T = 400$		
			h_c/h_f			h_c/h_f			h_c/h_f		
			8	6	4	8	6	4	8	6	4
1	S-S	FG	0.1595	0.1661	0.1779	0.1538	0.1607	0.1729	0.1487	0.1560	0.1686
	S-S	UD	0.1566	0.1617	0.1703	0.1509	0.1562	0.1651	0.1457	0.1513	0.1606
2	S-S	FG	0.6289	0.6549	0.7016	0.6086	0.6362	0.6849	0.5899	0.6194	0.6708
	S-S	UD	0.6180	0.6380	0.6721	0.5976	0.6191	0.6553	0.5785	0.6020	0.6408
3	S-S	FG	1.3837	1.4406	1.5427	1.3394	1.3998	1.5065	1.2980	1.3629	1.4756
	S-S	UD	1.3605	1.4047	1.4803	1.3159	1.3636	1.4440	1.2740	1.3262	1.4126
1	C-C	FG	0.3528	0.3673	0.3934	0.3414	0.3568	0.3840	0.3309	0.3474	0.3760
	C-C	UD	0.3467	0.3579	0.3770	0.3353	0.3473	0.3675	0.3245	0.3377	0.3594
2	C-C	FG	0.9483	0.9870	1.0564	0.9177	0.9587	1.0313	0.8891	0.9331	1.0097
	C-C	UD	0.9327	0.9628	1.0143	0.9019	0.9344	0.9892	0.8729	0.9084	0.9673
3	C-C	FG	1.8026	1.8741	2.0047	1.7425	1.8210	1.9575	1.6881	1.7710	1.9152
	C-C	UD	1.7744	1.8304	1.9290	1.7140	1.7772	1.8819	1.6579	1.7268	1.8394
1	C-S	FG	0.2474	0.2575	0.2759	0.2391	0.2499	0.2689	0.2316	0.2431	0.2631
	C-S	UD	0.2430	0.2508	0.2642	0.2347	0.2431	0.2572	0.2270	0.2361	0.2511
2	C-S	FG	0.7860	0.8183	0.8762	0.7607	0.7949	0.8555	0.7372	0.7740	0.8379
	C-S	UD	0.7727	0.7976	0.8404	0.7472	0.7741	0.8195	0.7234	0.7528	0.8015
3	C-S	FG	1.5962	1.6613	1.7781	1.5446	1.6136	1.7357	1.4961	1.5703	1.6992
	C-S	UD	1.5703	1.6212	1.7085	1.5184	1.5733	1.6661	1.4694	1.5295	1.6292

more flexible as the slenderness ratio grows. Regarding the inverse relation between the stiffness and flexibility, it will be perceived that the total stiffness decreases and this phenomenon causes a reduction in the natural frequency of the beam.

Finally, Table 11.8 and Figure 11.3 probe the mechanical reaction of the system to changes in the core-to-facesheet thickness ratio. As mentioned in the above paragraphs, it is clear that the frequency of the sandwich beam will be lessened as the temperature rise grows. Moreover, the C-C beams possess greater natural frequencies compared with C-S and S-S ones. Similar to previous trends, it can be concluded from Table 11.8 and Figure 11.3 that the dynamic response of the continuous system will decrease as the thickness ratio between the core and the face sheets increases. The major reason for this trend can be found by remembering that the stiffness of the nanocomposite face sheets is very big compared with that of the homogeneous core. Hence, improvement of the portion of the nanocomposite face sheets in the total thickness of the sandwich beam helps to enhance the system's equivalent stiffness. As expressed in the above paragraphs, the natural frequency has a direct relation with the stiffness; therefore, it is natural to see a decrease in the natural frequency

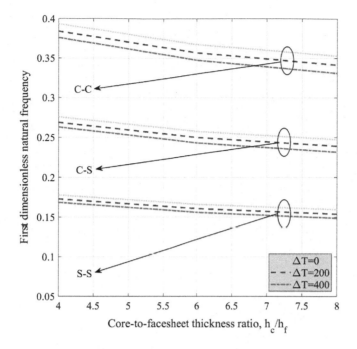

FIGURE 11.3 First natural frequency of C-C, S-S, and C-S sandwich beams with FG-CNTR nanocomposite facesheets with different core-to-facesheet thickness ratios.

once the ratio between the thickness of the core and face sheets is increased. In other words, the larger is the portion of the core in the total thickness, the lesser will be the stiffness of the beam.

11.3 CONCLUDING REMARKS

In this chapter, the frequency analysis of sandwich beams with two CNTR nanocomposite facesheets was produced on the basis of the implementation of the Timoshenko beam hypothesis in association with the dynamic form of the principle of virtual work. The impact of temperature on the frequency of the system was included by considering the axial thermal force applied on the beam as well as regarding for the temperature-dependency of the constituent material. A brief review of the most important results is presented below:

- It was shown that the frequency of the beam can be dramatically influenced by the volume fraction of the CNTs in the facesheets.
- It was shown that using a FG pattern for the distribution of the CNTs across the thickness can better increase the stiffness of the system than using the UD.
- In accordance with previous estimations, C-C beams were the best choice for the cases where the structure must be able to endure high frequencies, followed by C-S and S-S ones.

12 Thermally Affected Vibrational Responses of Multilayered FG-GPLR Nanocomposite Beams

12.1 PROBLEM DEFINITION

The present chapter investigates the free vibration problem of multilayered nanocomposite beams reinforced by dispersion of nanosize GPLs in a polymeric matrix. The effective material properties of the multilayered beam-type element will be developed using the Halpin-Tsai micromechanical method for the case of an N-layered nanocomposite structure. The governing equations of the beam will be derived on the basis of the expansion of the Reddy's TSDT using Hamilton's principle. These equations are presented in Eqs. (2.133)–(2.135). Afterward, the equations will be solved within the framework of Navier's method to satisfy the simply supported BC at the ends of the beam. Numerical examples reveal the fact that the natural frequency of the system can be critically influenced by changing the magnitude of the temperature rise. Also, the content of the nanosize reinforcing phase plays an indispensable role in determining the system's natural frequency.

12.2 NUMERICAL RESULTS AND DISCUSSION

In this section, numerical investigations illustrate the influences of various parameters in determining the natural frequency of the system. These results will be presented in the framework of a group of examples. In all of the figures, the thickness of the beam is considered to be $h = 2$ mm. Also, the beam is assumed to be manufactured from a 10-layer GPLR nanocomposite unless another value is reported. The material properties of the polymeric matrix and the reinforcing GPLs are assumed to be as same as those reported in Chapter 8. To simplify the process, the following dimensionless frequency will be utilized:

$$\Omega = \omega \frac{L^2}{h} \sqrt{\frac{\rho_m}{E_m}} \tag{12.1}$$

In Figure 12.1, variations of the dimensionless frequency versus weight fraction of the GPLs are drawn for multilayered FG nanocomposite beams, excluding the

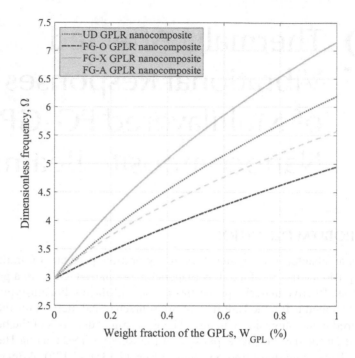

FIGURE 12.1 Variation of the dimensionless frequency of FG-GPLR nanocomposite beams versus weight fraction of the GPLs for various types of FG nanocomposite materials ($L/h = 10$, $\Delta T = 0$).

effects of thermal gradient. It is obvious that increasing the content of the GPLs in the continuous system results in an increase in the dynamic response of the system. The reason is the enhancement of the beam's stiffness when the composition of the nanocomposite material utilizes a larger amount of the reinforcements. Moreover, FG-X nanocomposite beams provide higher natural frequencies compared with all other types of GPLR nanocomposites; thus, the act of reinforcement produces better results than other types of nanocomposite beams. Among all types of FG nanocomposites, the smallest response belongs to FG-O GPLR nanocomposites.

Investigating the effect of a temperature rise on the dimensionless natural frequency of the beam is the subject of Figure 12.2. According to this figure, the dimensionless frequency of the beam will be reduced when a greater value is assigned to the temperature rise. The slope of this reduction is roughly the same for all types of FG nanocomposites. Again, it can be seen that the peak natural frequency in any desired temperature rise corresponds with implementation of FG-X GPLR nanocomposite beams followed by UD GPLR, FG-A GPLR, and FG-O GPLR, respectively. Note that if the aim of applying thermal gradient is to damp the dynamic response of the nanocomposite system, a huge temperature gradient must be applied. Looking at the figure, a 200 K temperature rise cannot

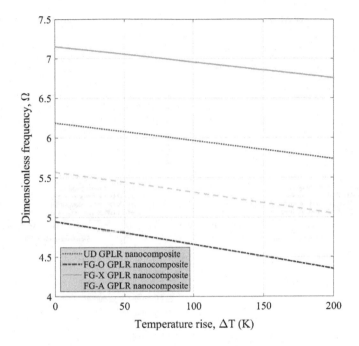

FIGURE 12.2   Variation of the dimensionless frequency of FG-GPLR nanocomposite beams versus temperature rise for various types of FG nanocomposite materials ($L/h = 10$, $W_{GPL} = 1\%$).

damp the oscillation of the continua and this is the result of adding the nanosize GPLs to the system.

The next figure (i.e., Figure 12.3) probes the effect of the slenderness ratio of the beam as a geometrical term on the natural frequency of the beam whenever the effects of temperature rise are considered. When thermal effects are excluded the dimensionless frequency of the GPLR nanocomposite beam will experience an initial increase followed by a stationary state. Indeed, the relation between the dimensionless frequency and slenderness ratio makes it possible to see an improvement in the dynamic response of the system when the slenderness ratio is amplified. However, this trend cannot be observed when a thermal gradient is included in the numerical investigation. Indeed, the stiffness-softening impact of the temperature rise on the equivalent stiffness of the system results in a critical reduction in the dynamic response of the system such that in some cases the dynamic response can be diminished completely. As with previous examples, the dynamic response of the system is maximized when FG-X GPLR nanocomposite beams are used.

At the end of this section, the influence of the number of layers on the dynamic response of the system will be investigated within the framework of

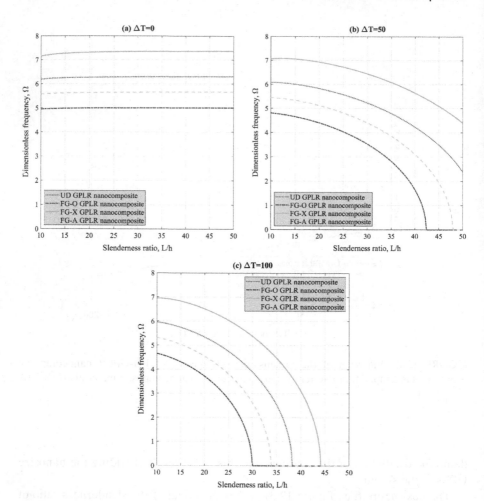

FIGURE 12.3 Variation of the dimensionless frequency of FG-GPLR nanocomposite beams versus slenderness ratio for various types of FG nanocomposite materials regarding for thermal effects ($W_{GPL} = 1\%$).

Figure 12.4. Based upon this figure, it can be seen that the dimensionless frequency of FG-X GPLR nanocomposite beams can be increased once the number of the layers is changed from five to fifteen. After this number of layers, no remarkable change in the dynamic response of the system can be observed. As expected, the dynamic response of the UD GPLR beams will not be affected by the number of layers. On the other hand, the dynamic response of the FG-O and FG-A GPLR nanocomposite beams will be lessened once a higher number of layers is utilized. Similar to FG-X GPLR nanocomposite structures, in FG-O and FG-A structures the effect of the layers' number cannot be seen clearly after the total number of layers exceeds roughly fifteen.

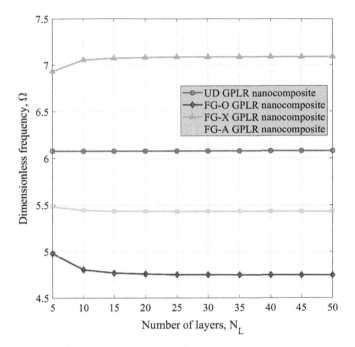

FIGURE 12.4   Variation of the dimensionless frequency of FG-GPLR nanocomposite beams versus number of layers for various types of FG nanocomposite materials ($W_{GPL} = 1\%$, $\Delta T = 50$).

12.3   CONCLUDING REMARKS

This chapter was arranged to show the effect of the thermal gradient and number of the beam's layers on the dynamic responses of the nanocomposite beams reinforced via dispersion of GPLs in the polymeric resin. The homogenization procedure was accomplished based on the Halpin-Tsai method for the case of a multilayered sandwich nanocomposite structure. The governing equations of the problem were obtained by implementing the Reddy's beam hypothesis, also known as TSDT, in association with the dynamic form of the principle of virtual work. At the end, the achieved governing equations were solved using the Navier-type solution for the purpose of satisfying the simply supported BC at the ends of the beam. The most crucial findings of the above chapter are as follows:

- The natural frequency of the sandwich beam can be strengthened by increasing the content of the available nanosize reinforcements in the media.
- Among different types of FG nanocomposite beams, FG-X possesses the greatest response followed by UD, FG-A, and FG-O, respectively.
- Increasing the number of the layers does not make a significant change in the dynamic response of the system.
- Subjecting the nanocomposite media to a thermal environment results in the reduction of the equivalent stiffness of the structure and this issue will result in reaching lower natural frequencies.

13 Thermo-Magnetically Influenced Vibration Analysis of GPLR Nanocomposite Beams

13.1 PROBLEM DEFINITION

This chapter considers the mathematical modeling of the free vibration problem of GPLR nanocomposite beams resting on an elastic substrate given the influences of thermal gradient and magnetic field on the vibrational response of the system. The beam-type element is of length L and thickness h that is assumed to be rested on a Winkler-Pasternak elastic foundation. The GPLs are assumed to be dispersed in the polymeric matrix via UD pattern and the equivalent thermo-mechanical material properties of the constituent materials are developed using the Halpin-Tsai micromechanical method, as explained in Section 2.3.2. The motion equations of the problem will be derived based on expanding Hamilton's principle for refined shear deformable beams. Because of the effects of magnetic field on the dynamic responses of the nanocomposite structure, the magnetic induction relations of Maxwell must be written for the case of a beam-type element to reach the final version of the governing equations of the problem. It is worth mentioning that in the present chapter, only UTR-type thermal loading will be discussed. The governing equations of the problem can be found by referring to Eqs. (2.163)–(2.165). In addition to the above explanations, the magnetic field is assumed to be uniform and it will be applied in the longitudinal direction, parallel with the beam's axis. The governing equations will be solved using Galerkin's method, a powerful solution that can cover the effects of various types of BCs on the mechanical response of the continuous systems. It will be shown below that increasing the intensity of the magnetic field can help to improve the frequency range that can be supported by the nanocomposite beam. Also, thermal losses can play a crucial role in reducing the thermo-magnetically affected dynamic responses of the system.

13.2 NUMERICAL RESULTS AND DISCUSSION

This section investigates the impacts of various terms on the mechanical response of the continuous systems manufactured from UD GPLR nanocomposite materials. The thermo-mechanical material properties of the reinforcing phase (i.e., GPL) and host matrix (i.e., PMMA) can be considered to be as same as those reported in the

TABLE 13.1

Comparison of Dimensionless Natural Frequencies of CNTR Nanocomposite Beams without Foundation ($L/h = 15$, $V_{CNT} = 0.12$)

Distribution Pattern	Present	Wattanasakulpong and Ungbhakorn (2013)
FG-O CNTR nanocomposite	0.7455	0.7527
UD CNTR nanocomposite	0.9745	0.9753
FG-X CNTR nanocomposite	1.1152	1.1150

previous chapter of this book. The magnetic permeability is $\eta = 4\pi \times 10^{-7}$. Also, in the following examples the thickness of the structure is assumed to be $h = 2$ mm. To show the efficiency of the presented method, Table 13.1 presents the fundamental natural frequencies of FG-CNTR nanocomposite beams in comparison to the literature. Based on this table, it can be certified that the dynamic responses of nanocomposite beams can be estimated via presented modeling with a remarkable precision. Indeed, the results of our method are in reliable agreement with those reported by former researches in the literature.

In the following numerical examples, the dimensionless forms of the natural frequency and foundation coefficients are utilized for the sake of simplicity. The dimensionless forms can be found as follows:

$$\Omega = \omega \frac{L^2}{h} \sqrt{\frac{\rho_m}{E_m}}, \quad K_w = k_w \frac{L^4}{D^*}, \quad K_p = k_p \frac{L^2}{D^*}, \quad D^* = \frac{E_m h^3}{12(1 - v_m^2)} \quad (13.1)$$

Focusing on the role of the geometrical parameters on the thermo-elastic responses of the nanocomposite system, the variation of the dimensionless frequency versus slenderness ratio of the beam is plotted in Figure 13.1 for various types of edge supports at the ends of the beam-type element. In this diagram, the temperature rise is tuned to find the effect of this term on the dynamic response of the system. It can be seen that increasing the temperature rise results in a reduction in the dimensionless frequency of the vibrating system. The reason for this is the softening influence of the temperature rise on the equivalent stiffness of the beam. Therefore, the cross-sectional rigidities of the beam will be decreased and thus, the natural frequency will be lessened. In this figure, the effect of changing the BC on the natural frequency of the nanocomposite beam can be observed. In other words, the beams with C-C BCs can tolerate greater dynamic excitations among other types of BCs followed by C-S and S-S ones. Note that the effects of elastic medium and magnetic field are dismissed.

Figure 13.2 monitors the variation of the natural frequency of the beam versus weight fraction of the GPLs for various amounts of temperature rise once the BCs of the beam are changed. Clearly, the vibrational response of the system will be enhanced when the content of the nanosize GPLs in the composition of the GPLR

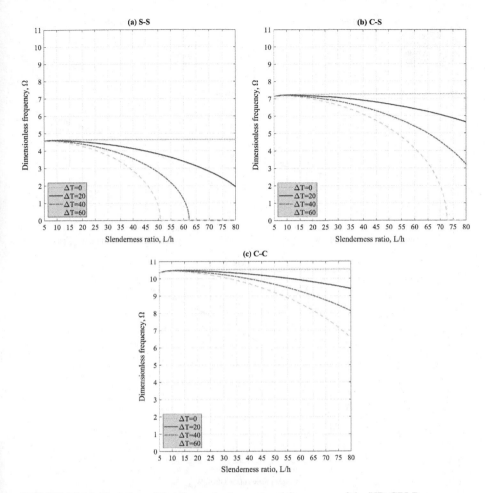

FIGURE 13.1 Variation of the dimensionless natural frequency of the UD GPLR nanocomposite beams versus slenderness ratio for various temperature rises regarding the effect of BCs on the thermo-mechanical response ($W_{GPL} = 0.5\%$, $K_w = 0$, $K_p = 0$, $H = 0$).

nanocomposite is increased. The reason for this trend is the amplifying role of the GPLs on the increase of the system's effective stiffness because of the superior elastic features of GPLs compared with macroscale materials. Again, increasing the temperature gradient decreases the natural frequency due to the softening influence of the temperature on the stiffness of the system. Similar to the former example, the dimensionless frequency of the system possesses its maximum magnitude when the C-C BC is chosen compared with the case of C-S and S-S ones. In fact, the structural stiffness of the C-type edge support is greater than that of the S-type one.

The next example shows how the natural frequency of the GPLR nanocomposite beam can be influenced by changing the intensity of the magnetic field. To

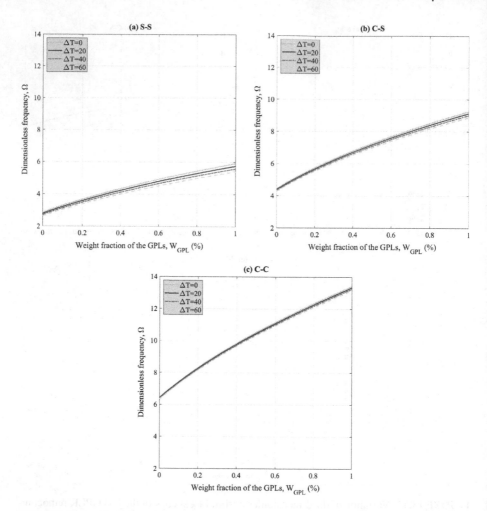

FIGURE 13.2 Variation of the dimensionless natural frequency of the UD GPLR nanocomposite beams versus weight fraction of the GPLs for various temperature rises regarding the effect of BCs on the thermo-mechanical response ($L/h = 20$, $K_w = 0$, $K_p = 0$, $H = 0$).

this purpose, a GPLR S-S beam is implemented in Figure 13.3 to draw the variation of the dimensionless frequency of the beam versus magnetic field intensity for various amounts of temperature rise. According to this diagram, it is easy to reach high-magnitude natural frequencies once the intensity of the magnetic field is increased. It is clear that the variations are critically nonlinear such that a small increment in the magnetic field intensity will be resulted in a noticeable increase in the value of the dimensionless frequency. Again, the influences of foundation parameters are dismissed and only thermo-magnetically influenced analysis is presented. It is important to point out that the impact of the temperature gradient on the dynamic response of the system will be attenuated as the intensity of the applied magnetic field is reached at approximately $H = 0.05 \times 10^8$ A/m. Therefore,

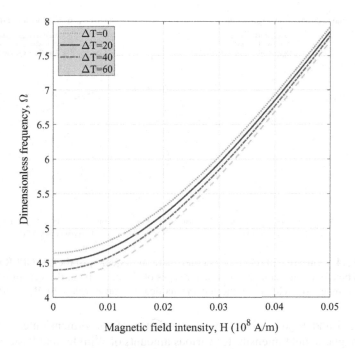

FIGURE 13.3 Variation of the dimensionless natural frequency of the UD GPLR nanocomposite S-S beams versus magnetic field intensity for various temperature rises regarding the effect of BCs on the thermo-mechanical response ($L/h = 20$, $W_{GPL} = 0.5\%$, $K_w = 0$, $K_p = 0$).

one of the ways of defeating the thermal losses is to implement high-range magnetic field in the environment.

In another example, the combined influences of foundation type, thermal gradient, and existence or absence of the magnetic field on the frequency response of the GPLR nanocomposite beam are surveyed within the framework of Figure 13.4. In this figure, the variation of the dimensionless natural frequency versus temperature rise is plotted for various types of elastic seats of the structure in both presence and absence of magnetic field. Clearly, the frequency of the beam will go through a decreasing path as the temperature rise becomes greater. The reason for this phenomenon was mentioned above. However, it can be easily seen that the beams rested on two-parameter elastic substrate can endure higher natural frequencies followed by those rested on Winkler substrate and those without any seat, respectively. In fact, the structural stiffness of the element will increase when the stiffness of the foundation's springs increases. Therefore, it is natural to see such an increase in the frequency of the nanocomposite beam due to the direct relation between the dynamic response and system's stiffness. In addition to the aforementioned phenomenological trends, one should pay attention to the increasing influence of the magnetic field on the natural frequency. As seen in this figure and previous one, the induced Lorentz force acts in a way that results in amplification of the dynamic response of the nanocomposite structure.

The next numerical examples show the qualitative impacts of the foundation parameters on the frequency of the nanocomposite beams. To this purpose,

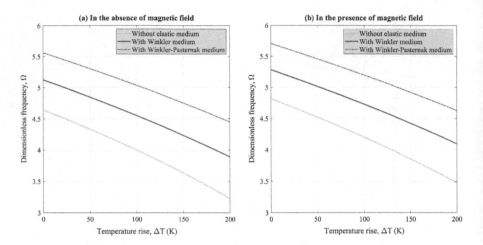

FIGURE 13.4 Variation of the dimensionless natural frequency of the UD GPLR nanocomposite S-S beams versus temperature rise for types of elastic seats for the structure regarding the effect of magnetic field on the thermo-mechanical response ($L/h = 20$, $W_{GPL} = 0.5\%$).

Figure 13.5 and Figure 13.6 show the variation of the dimensionless frequency against magnetic field intensity for various amounts of Winkler and Pasternak coefficients, respectively. In these figures, the intensifying effect of the magnetic field intensity on the frequency of the system can be seen as same as Figure 13.3. It can be seen that adding each of the Winkler or Pasternak coefficients will be resulted in an increase in the value of the system's frequency. The reason for this trend is the stiffness enhancement obtained by increasing the stiffness of the foundation parameters. However, the main object of these diagrams is to present a numerical comparison

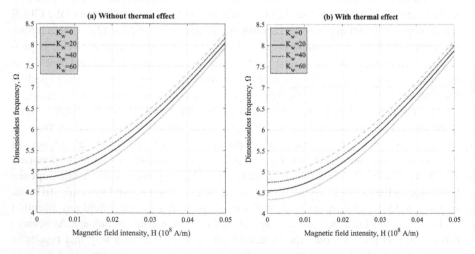

FIGURE 13.5 Variation of the dimensionless natural frequency of the UD GPLR nanocomposite S-S beams versus magnetic field intensity for different Winkler coefficients regarding the effect of temperature rise on the thermo-mechanical response ($L/h = 20$, $W_{GPL} = 0.5\%$, $K_p = 0$).

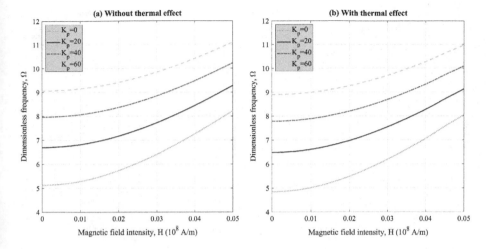

FIGURE 13.6 Variation of the dimensionless natural frequency of the UD GPLR nanocomposite S-S beams versus magnetic field intensity for different Pasternak coefficients regarding the effect of temperature rise on the thermo-mechanical response ($L/h = 20$, $W_{GPL} = 0.5\%$, $K_w = 50$).

between the effects of these two terms on the dynamic response of the system. Obviously, using the same dimensionless coefficients for each of these coefficients results in different amplifications in the dimensionless frequency of the nanocomposite beam. It is clear that the Pasternak coefficient can increase the frequency of the system many times more than the Winkler one. The effects of temperature rise on the dimensionless response once each of the Winkler or Pasternak coefficients are varied are also shown in these figures. It is clear that the effect of the thermal gradient on the dynamic response can be better seen in the case of changing the Winkler coefficient. This observation reveals that the Winkler coefficient can resist thermal losses less than the Pasternak one, which certifies that the Pasternak coefficient is more able to improve the dimensionless frequency of the system.

As the final example in this chapter, Figure 13.7 presents the variation of the dimensionless frequency versus temperature rise for various types of BCs for two types of beam-type elements. It can be easily seen that increasing the temperature rise will result in a decrease in the dimensionless frequency of the system for all types of BCs. Moreover, the effect of the BC can be seen, too. C-C beams can provide greater dimensionless frequencies compared with other types of beams with C-S and S-S BCs, respectively. The most significant highlight of the figure is the influence of increasing the content of the nanosize GPLs to the polymeric matrix on the dimensionless response of the beam. Adding a small amount of the GPLs can result in a remarkable increase in the range of the natural frequencies that can be supported by the beam. The superiority of the GPLR nanocomposite materials to the polymers can be observed in this figure. The reason of this remarkable difference in the range of the frequencies which can be tolerated is the marvelous stiffness improvement that can be induced in the polymers by adding a small amount of the nanosize reinforcements to them.

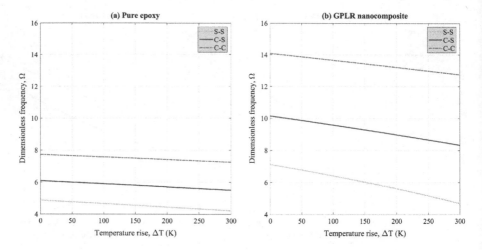

FIGURE 13.7 Variation of the dimensionless natural frequency of the UD GPLR nanocomposite beams versus temperature rise for different BCs regarding the effect of GPLs' weight fraction on the thermo-mechanical response ($L/h = 20$, $K_w = 50$, $K_p = 10$, $H = 10^6$ A/m).

13.3 CONCLUDING REMARKS

This chapter was concerned with investigating the thermo-magnetically affected vibrational characteristics of GPLR nanocomposite beams resting on elastic foundation. In this problem, the beam was presumed to be manufactured from UD GPLR nanocomposite material and it was placed in an environment that was influenced by existence of magnetic field. The equivalent material properties of the beam were developed based on the Halpin-Tsai micromechanical procedure for the case of a single-layered beam-type element. Moreover, the mathematical representation of the beam's motion was developed using the concept of the dynamic form of the principle of virtual work incorporated with a refined three-variable higher-order beam theorem to include the effects of shear deflection in the procured dynamic analysis. The effects of the existence of a magnetic field on the mechanical behavior of the continuous system was covered by extending Maxwell's induction relation for the structure. The following points review the results to emphasize the significance of the impacts of the involved variants on the mechanical characteristics of the beam:

- The dimensionless frequency of the beam can be remarkably increased by adding a small amount of the GPLs to the host polymer.
- The system's frequency can be dramatically increased by increasing the intensity of the applied uniform-type magnetic field in any desired content of the reinforcing GPLs.
- Once the attenuation of the system's oscillation is determined, applying a thermal gradient can be a useful alternative.
- Increasing the elastic coefficients of the medium will enlarge the dimensionless natural frequency. The effect of the Pasternak coefficient is many times greater than that of the Winkler coefficient.
- C-C nanocomposite structures reveal greater natural frequencies compared to C-S and S-S ones.

14 Damped Vibration Analysis of Multilayered FG-GPLR Nanocomposite Beams Embedded on Viscoelastic Substrate

14.1 PROBLEM DEFINITION

This chapter investigates the combined impacts of the foundation's parameters, distribution pattern of the GPLs, and content of the reinforcing phase on the natural frequency of the multilayered beam-type elements resting on a visco-Pasternak medium. The equivalent material properties of each of the beam's layers will be obtained using the micromechanical procedure expressed in Part 1 (see Sections 2.3.2 and 2.3.4). The kinematic relations of the beam-type elements used in the analysis can be derived based on Reddy's TSDT of beams. After extracting suitable relations for the components of valid strains in terms of the displacement field of the problem, the governing equations will be derived by expanding the dynamic form of the principle of virtual work for the beams simulated with the beam hypothesis. The coupled set of the governing equations can be found in Eqs. (2.133)–(2.135). In the following section, a numerical illustrative study will be provided to discuss the influences of various terms on the damped dynamic behaviors of GPLR nanocomposite beams. It can be seen that increasing the content of the GPLs available in the polymeric matrix results in a remarkable increase in the dimensionless frequency of the system. Also, the dynamic response of the system can be easily damped using a greater damping coefficient for the visco-Pasternak medium.

14.2 NUMERICAL RESULTS AND DISCUSSION

In this section, a set of numerical examples will be presented to show the effects of different parameters on the mechanical behavior of the FG-GPLR nanocomposite beams. The material properties of the GPLs and matrix are chosen to match those reported in the former chapters (e.g., see Chapter 8) concerning with the mechanical responses of GPLR nanocomposite structures. Herein, the results will be presented in the dimensionless form for the sake of simplicity. Thus, it is necessary to become familiar with the dimensionless form of the foundation parameters and frequency of the structure before discussing about the phenomenological trends:

$$\Omega = \omega \frac{L^2}{h} \sqrt{\frac{\rho_m}{E_m}}, \ K_w = k_w \frac{L^4}{D^*}, \ K_p = k_p \frac{L^2}{D^*}, \ C_d = \frac{c_d L}{\rho_m D^*}, \ D^* = \frac{E_m h^3}{12\left(1-v_m^2\right)} \quad (14.1)$$

where K_w, K_p, and C_d are the dimensionless forms of the Winkler, Pasternak, and damping coefficients of the viscoelastic medium, respectively.

As the first example, the variation of the dimensionless frequency of both polymer and UD GPLR nanocomposite beams against temperature rise is plotted in Figure 14.1 for various amounts of foundation parameters. It can be observed that adding a low content of the GPLs to the polymer matrix can improve the natural frequency of the system in a remarkable manner. This is due to the outstanding stiffness of the GPLs which results in an excellent equivalent stiffness for the nanocomposite material. Also, this figure reveals that adding the stiff coefficients of the viscoelastic substrate can enhance the dynamic endurance of the continuous system. The dimensionless frequency of the system can be better damped whenever the structure is rested on a viscoelastic medium with a high damping coefficient. Therefore, there is a correlation between thermal and viscose losses in the figure.

Figure 14.2 shows the variation of the dimensionless frequency of the GPLR nanocomposite beams versus the damping coefficient of the visco-Pasternak substrate for different amounts of GPLs' weight fraction in order to put an emphasis on the significance of the role of weight fraction of the nanosize reinforcements on the determination of the natural frequency of various types of FG-GPLR nanocomposite structures. It can be seen that eventually the increase of the foundation's damping coefficient results in complete attenuation of the natural frequency. Based on this figure, the dynamic response of the continuous system will be increased whenever the weight fraction of the GPLs is increased. The reason for this is the enhancement of the system's effective stiffness in the case of high weight fractions of the GPLs. Among various distribution patterns for the GPLs across the thickness of the beam, the FG-X GPLR nanocomposites show greater natural frequencies in comparison with the other types, followed by UD GPLR, FG-V GPLR, and FG-O GPLR ones, respectively. It may be interesting to point out that the amount of the damping coefficient required to completely damp the frequency of the system differs for various types of GPLR nanocomposites. In other words, the frequency of FG-X GPLR nanocomposite beams will dissipate later than other types of nanocomposite structures, followed by UD GPLR, FG-V GPLR, and FG-O GPLR ones, respectively. This trend emphasizes the greater stiffness of the FG-X GPLR nanocomposite materials once again.

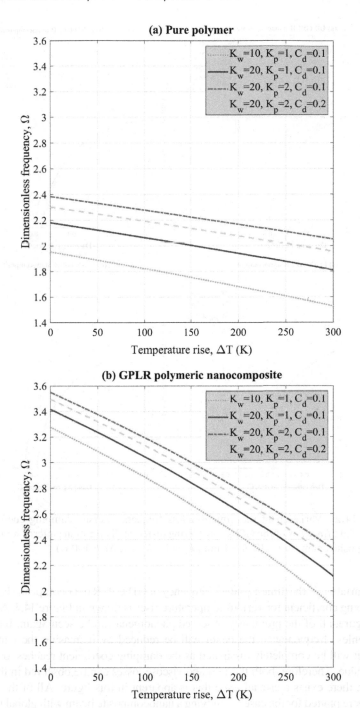

FIGURE 14.1 Variation of the dimensionless frequency versus temperature rise for various coefficients of the viscoelastic foundation considering both pure polymer and GPLR nanocomposite as the constituent material of the structure ($L/h = 10$, $h = 1$ mm).

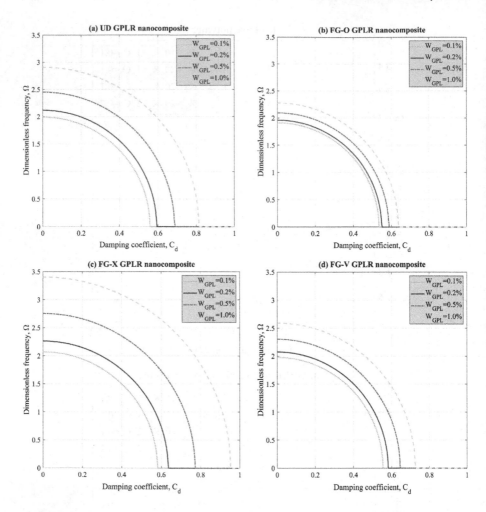

FIGURE 14.2 Variation of the dimensionless frequency versus damping coefficient for various weight fractions of the GPLs considering different distribution patterns for the GPLR nanocomposite beam ($L/h = 10$, $h = 1$ mm, $K_w = 10$, $K_p = 1$, $\Delta T = 100$ K).

The variation of the dimensionless frequency of UD GPLR nanocomposite beams versus damping coefficient for various temperature rises is drawn in Figure 14.3. According to this figure, all of the previously observed phenomena can be seen again. Indeed, the dimensionless frequency of the beam will be reduced by increasing the temperature rise and it will be completely dissipated as the damping coefficient reaches to a certain critical value. Therefore, both thermal and viscose losses can be observed in the figure. However, there exists a crucial issue that is apparent in this figure. All of the previous figures were plotted for the case of studying a nanocomposite beam with global thickness of 1 mm; whereas, this figure investigates the damped vibration responses of beams of thickness 2 mm. It can be seen that the damping coefficient required to completely damp the fluctuation of the system will grow as the thickness of the structure will be increased.

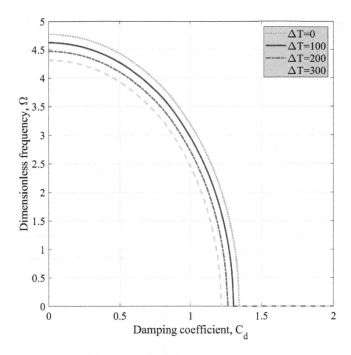

FIGURE 14.3 Variation of the dimensionless frequency of UD GPLR nanocomposite beams versus damping coefficient for various amounts of temperature rise ($L/h = 10$, $h = 2$ mm, $W_{GPL} = 0.5\%$, $K_w = 10$, $K_p = 1$).

Finally, the variation of the dimensionless frequency versus slenderness ratio of the beam is plotted in Figure 14.4 for various contents of the GPLs in the composition of the GPLR nanocomposite and also considering the thermal losses and type of substrate. It can be seen that increasing the foundation's viscose term or thermal gradient give the same result. In the absence of the damping coefficient of the medium and temperature rise, the dimensionless frequency of the beam will increase up to a certain limited value and afterward, remains unchangeable. However, such a trend cannot be observed in the case of considering either damping effect or thermal effect or both. In the latter situations, the dimensionless frequency will be reduced due to the losses caused by thermal gradient and damping phenomenon.

14.3   CONCLUDING REMARKS

In this chapter the thermo-viscoelastically affected free vibration problem of a multilayered FG-GPLR nanocomposite beam was solved for S-S beams using the TSDT of beams. The equivalent homogenized properties of the nanocomposite were developed according to the Halpin-Tsai micromechanical procedure for multilayered structures. The governing equations of the problem were derived based upon the dynamic form of the principle of virtual work in association with Reddy's TSDT. The beam was presumed to be subjected to a uniform-type temperature gradient

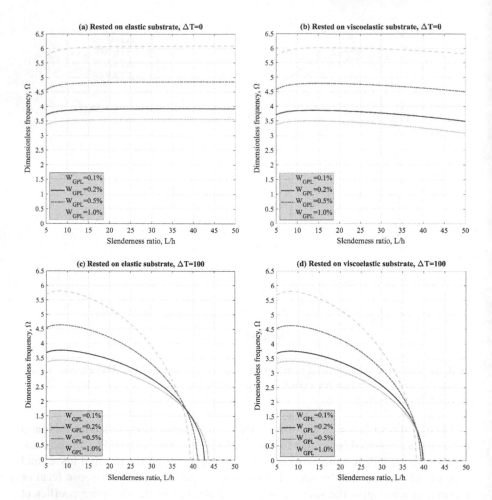

FIGURE 14.4 Variation of the dimensionless frequency of UD GPLR nanocomposite beams versus slenderness ratio for various amounts of GPLs' weight fraction considering the effects of both thermal and viscose losses ($L/h = 10$, $h = 2$ mm, $K_w = 10$, $K_p = 1$).

and it rested on a three-parameter visco-Pasternak medium. The highlights of this chapter are presented below:

- The dynamic response of the GPLR nanocomposite structures can be improved by increasing the content of the GPLs in the media.
- Among different types of FG nanocomposites, FG-X GPLR nanocomposites can provide greater natural frequencies compared with the other types, followed by UD GPLR, FG-V GPLR, and FG-O GPLR ones, respectively.
- Increasing the elastic stiffnesses of the foundation's springs can be one of the best ways to intensify the dimensionless frequency of the nanocomposite beam.
- The dimensionless frequency of the GPLR nanocomposite beam can be easily dissipated by increasing either the thermal gradient or the damping coefficient of the viscoelastic substrate.

15 Influence of External Non-Uniform Magnetic Field on the Natural Frequency of FG-GOR Nanocomposite Beams

15.1 PROBLEM DEFINITION

This chapter investigates the effect of a non-uniform external magnetic field on the vibrational characteristics of polymeric nanocomposite beams reinforced with GOs dispersed in the matrix. The effective material properties of the nanocomposite material will be derived using the Halpin-Tsai micromechanical scheme, as explained in Section 2.3.3. The motion equations of the problem can be derived on the basis of Hamilton's principle incorporating the displacement field of the refined shear deformable beam theory. According to the implemented shear deformable beam model, the strain–displacement relations of the beam will be achieved such that the influences of shear stress and strain are covered. In this study there exists a non-uniform magnetic field that affects the motion equations of the problem. This external field possesses a trigonometric form and varies in various spatial coordinates along the axis of the beam. To include the impact of such a non-uniform field, the magnetic induction relations of Maxwell must be expanded for the refined higher-order beams. Following this procedure, the final governing equations of the problem will be as same as those reported in Eqs. (166)-(168). Once the governing equations are solved via Galerkin's method, the results of the problem can be extracted for various types of BCs. The results of this chapter will be presented in the following section.

15.2 NUMERICAL RESULTS AND DISCUSSION

In this section some numerical illustrations will be presented to show the effects of various variants on the dynamic response of GOR nanocomposite beams. The material properties of the nanosize GOs and polymeric matrix can be assumed to be similar to those reported in Zhang et al. (2018). The thickness of the beam is $h = 2$ mm in all of the following illustrations. Also, the magnetic permeability is $\eta = 4\pi \times 10^{-7}$.

TABLE 15.1
Comparison of the First Dimensionless Frequency of S-S CNTR Nanocomposite Beams (L/h = 15, V_{CNT} = 0.12)

	Distribution Type		
Source	UD CNTR	FG-O CNTR	FG-X CNTR
Wattanasakulpong and Ungbhakorn (2013)	0.9976	0.7628	1.1485
Zhang et al. (2018)	0.9842	0.7595	1.1249
Present model	0.9904	0.7528	1.1399

The following dimensionless forms of natural frequency and magnetic field intensity are used for the sake of simplicity:

$$\Omega = \omega \frac{L^2}{h} \sqrt{\frac{\rho_m}{E_m}}, \quad H_0 = \frac{\eta A L^2}{E_{GO} I} H_x^2, \quad I = \frac{bh^3}{12} \tag{15.1}$$

Table 15.1 and Table 15.2 are presented to show that the presented methodology can satisfy a reliable precision while estimating the mechanical response of the nanocomposite beams in this model. As tabulated in these tables, the dimensionless frequency amounts achieved from our model are in a notable agreement with those reported by Wattanasakulpong and Ungbhakorn (2013) and also Zhang et al. (2018) for S-S nanocomposite beams.

The influence of the non-uniform magnetic field on the dimensionless frequency of UD GOR nanocomposite beams can be observed in Figure 15.1. In this figure, the variation of the dimensionless frequency versus slenderness ratio is plotted as the intensity of the applied magnetic field is varied. It can be seen that increasing the

TABLE 15.2
Comparison of the First Dimensionless Frequency of S-S GOR Nanocomposite Beams (W_{GO} = 0.3%)

		Distribution Type		
L/h	Source	FG-X GOR	FG-O GOR	UD GOR
10	Zhang et al. (2018)	0.3379	0.2921	0.3159
	Present model	0.3576	0.3013	0.3095
15	Zhang et al. (2018)	0.2271	0.1959	0.2121
	Present model	0.2411	0.2009	0.2079
20	Zhang et al. (2018)	0.1708	0.1473	0.1595
	Present model	0.1815	0.1461	0.1564

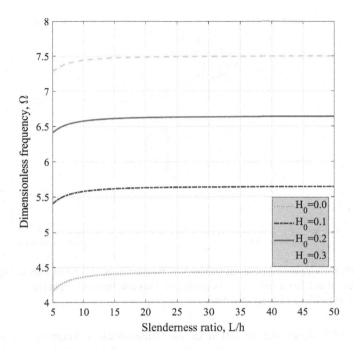

FIGURE 15.1 Variation of the dimensionless frequency of UD GOR nanocomposite S-S beams versus slenderness ratio of the beam for various amounts of the magnetic field intensity ($W_{GO} = 1\%$).

slenderness ratio results in an initial increase in the dimensionless frequency after which the frequency remains constant. The reason of this phenomenon is the direct relation between the dimensionless form of the natural frequency and the structure's slenderness ratio. The most crucial highlight of this diagram is the amplifying effect of the magnetic field on the dimensionless response of the system. It can be clearly seen that increasing the magnitude of the magnetic field causes an increase in the mechanical response of the system. The induced magnetic force plays an important role in enhancing the system's natural frequency because it improves the equivalent structural stiffness of the nanocomposite structure.

Figure 15.2 plots the variation of the dimensionless frequency of the nanocomposite beams versus slenderness ratio once the weight fraction of the GOs is changed. In this figure, two cases of existence and absence of magnetic field are included. It is obvious that increasing the magnetic field intensity is one way to increase the system's dynamic response. This increase occurs in such a way that the minimum response in the case of applying magnetic field will be greater than the maximum response in the cases that the magnetic field is absent. Due to the outstanding stiffness of nanosize GOs, the natural frequency of the beam experiences a big increase when a small amount of them is added to the composition of the GOR nanocomposite. Hence, achieving great natural frequencies is possible by tailoring the content of the GOs in the polymer matrix.

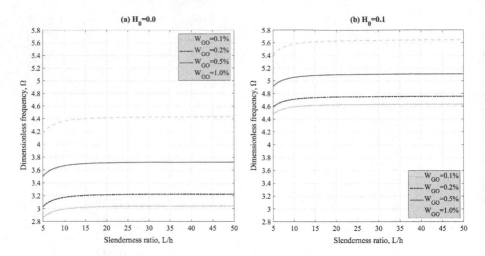

FIGURE 15.2 Variation of the dimensionless frequency of UD GOR nanocomposite S-S beams versus slenderness ratio of the beam for various amounts of weight fraction of the GOs considering the effects of magnetic field.

Figure 15.3 shows the variation of the dimensionless frequency versus GOs' weight fraction for various amounts of the magnetic field intensity for both S-S and C-C BCs. It can easily be seen that increasing the content of the GOs will result in an increase in the dimensionless frequency of the beam with a nonlinear slope. This phenomenon relates to the stiffness improvement that can be seen in the beam because of adding the amount of the GOs. Again it can be observed that increasing the intensity of the magnetic field results in an increase in the dimensionless

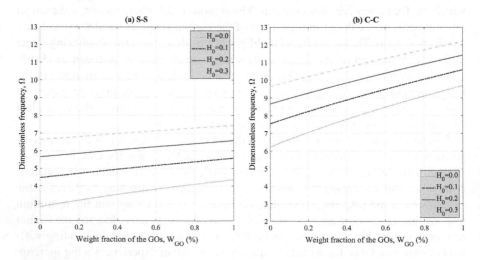

FIGURE 15.3 Variation of the dimensionless frequency of UD GOR nanocomposite beams versus GOs' weight fraction for various amounts of magnetic field intensity considering the effects of BCs ($L/h = 10$).

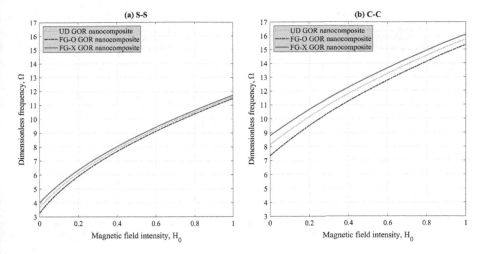

FIGURE 15.4 Variation of the dimensionless frequency of FG-GOR nanocomposite beams versus magnetic field intensity for various types of FG nanocomposites considering the effects of BCs ($L/h = 10$, $W_{GO} = 0.5\%$).

frequency. This observation was seen in the former figures, too. In addition to the abovementioned amplifying impacts, it is shown that C-C nanocomposite beams can support higher natural frequencies compared with S-S ones. The reason of this is the greater geometrical stiffness of the C-C beams in comparison with the S-S ones. Hence, C-C GOR nanocomposite structures placed in a magnetic environment can be employed in applications concerned with huge dynamic fluctuations.

The variation of the dimensionless frequency versus magnetic field intensity is plotted in Figure 15.4 for various types of FG nanocomposites. It can be observed that all of the previous results can be seen in this figure, too. In fact, the dimensionless frequency of the system can be enhanced by selecting C-C beams subjected to a high-intensity magnetic field rather than S-S ones placed in the same environment. It is obvious that FG-X GOR nanocomposite beams can provide higher natural frequencies among all types of FG nanocomposites. This is the result of the greater cross-sectional rigidities of such nanocomposite beams in comparison with other types of FG nanocomposite structures. In other words, because the concentration of the GOs is in the top and bottom edges of the beam, the edges will be stiffer than other parts of the structure, which leads to higher cross-sectional rigidities for such nanocomposite structures compared with UD GOR and FG-O GOR ones. Generally, in FG-X nanocomposite beams, axes at a greater distance from the neutral axis will possess greater stiffness, too. Thus, such nanocomposites can reveal bigger mechanical responses, followed by UD and FG-O ones, respectively.

Finally, the variation of the dimensionless frequency versus the weight fraction of the GOs for various types of FG nanocomposites is shown in Figure 15.5 considering both S-S and C-C nanocomposite beams. According to this figure, it can be seen that increasing the content of the reinforcing GOs in the composition of the constituent material can improve the frequency of the structure by enhancing the system's equivalent stiffness. However, the amount of this reinforcement differs for various

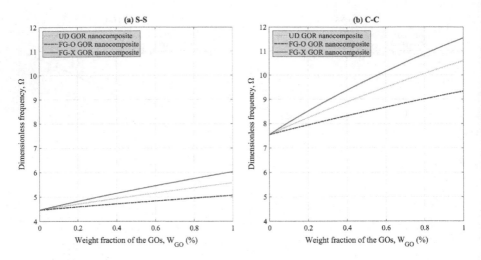

FIGURE 15.5 Variation of the dimensionless frequency of FG-GOR nanocomposite beams versus weight fraction of the GOs for various types of FG nanocomposites considering the effects of BCs ($L/h = 10$, $H_0 = 0.1$).

types of FG nanocomposites. In other words, the dynamic response of FG-X GOR nanocomposite beams can be enhanced by increasing the content of the GOs more than all other types of nanocomposite beams. Also, it is shown that C-C beams support higher natural frequencies in comparison with S-S ones because of the greater geometrical stiffness of such beams.

15.3 CONCLUDING REMARKS

In this chapter, the free vibration analysis of FG-GOR nanocomposite beams was carried out once the structure was subjected to a non-uniform magnetic field. The homogenization process was completed by implementing a micromechanical scheme for GOR polymeric nanocomposites with FG patterns for reinforcing the primary matrix. Thereafter, the set of PDEs controlling the motion of the beam were obtained according to the expansion of a refined higher-order beam model incorporated with the dynamic version of the principle of virtual work. The effects of magnetic field on the dynamic response of the continuous system were applied using Maxwell's magnetic induction relationships. At the end, the governing equations of the problem were solved in the framework of Galerkin's analytical method to extract the natural frequency of both S-S and C-C beams. The most significant highlights of this chapter follow:

- The natural frequency of the beam can be improved by increasing the content of the GOs in the constituent material.
- FG-X GOR nanocomposite beams provide greater frequencies compared with UD and FG-O ones.
- The dynamic response of the system will be increased as the magnetic field intensity increases.
- C-C nanocomposite beams can support greater dynamic responses compared with S-S ones.

16 Thermo-Mechanical Vibration Analysis of Multilayered FG-GPLR Nanocomposite Plates

16.1 PROBLEM DEFINITION

This chapter surveys the influence of thermal loading on the dynamic characteristics of sandwich FG-GPLR nanocomposite plate-type elements. To this purpose, the equivalent material properties of the nanocomposite will be developed using the Halpin-Tsai micromechanical scheme for multilayered nanocomposite structures. The derivation procedure can be found by referring to Sections 2.3.2 and 2.3.4. In addition, the mathematical representation of the vibration problem will be produced using the TSDT of the plates. The set of the governing equations can be found in the framework of Eqs. (2.191)–(2.195). Once the governing equations are developed they will be solved for the simply supported plates using the well-known Navier's analytical solution. The following sections carry out the numerical investigations then presenting the most important concluding remarks.

16.2 NUMERICAL RESULTS AND DISCUSSION

In this chapter, the impacts of different parameters on the vibrational responses of FG-GPLR multilayered nanocomposite plates will be studied in the framework of a group of illustrative examples. The total thickness of the plate-type element is $h = 2$ mm in all of the following examples. Also, the weight fraction of the reinforcing phase (i.e., GPLs) is 1% unless another value is reported. The material properties of the GPLs and polymeric matrix are identical to those presented in former chapters (e.g., Chapter 12). Before discussing the results of the study, it is best to prove the validity of the presented modeling in determination of the dynamic responses of FG nanocomposite plates. To this purpose, Table 16.1 depicts the comparison of the natural frequency responses of FG-GPLR nanocomposite plates with those obtained in research conducted by Song et al. (2017a). It can be seen that the modeling reveals accurate responses and shows a remarkable agreement between the results of this model with those reported in the open literature. The tiny differences between our results and those achieved by Song et al. (2017a) are a result of the different kinematic plate theorems that are utilized in these models; the results reached from our model are calculated implementing the TSDT while those reported in the aforementioned reference are calculated using the FSDT.

TABLE 16.1

Comparison of the Dimensionless Natural Frequencies of GPLR Nanocomposite Plates for Various Mode Numbers and Different Distribution Patterns

Pure Epoxy Plates

	Mode numbers (*m*, *n*)					
Source	(1, 1)	(2, 1)	(2, 2)	(3, 1)	(3, 2)	(3, 3)
Song et al. (2017a)	0.0584	0.1391	0.2132	0.2595	0.3251	0.4261
Present	0.0584	0.1391	0.2132	0.2595	0.3251	0.4261

UD GPLR Nanocomposite Plates

	Mode Numbers (*m*, *n*)					
Source	(1, 1)	(2, 1)	(2, 2)	(3, 1)	(3, 2)	(3, 3)
Song et al. (2017a)	0.1216	0.2895	0.4436	0.5400	0.6767	0.8869
Present	0.1216	0.2895	0.4436	0.5400	0.6767	0.8869

FG-O GPLR nanocomposite plates

	Mode Numbers (*m*, *n*)					
Source	(1, 1)	(2, 1)	(2, 2)	(3, 1)	(3, 2)	(3, 3)
Song et al. (2017a)	0.1020	0.2456	0.3796	0.4645	0.5860	0.7755
Present	0.0976	0.2353	0.3644	0.4464	0.5640	0.7479

FG-X GPLR nanocomposite plates

	Mode Numbers (*m*, *n*)					
Source	(1, 1)	(2, 1)	(2, 2)	(3, 1)	(3, 2)	(3, 3)
Song et al. (2017a)	0.1378	0.3249	0.4939	0.5984	0.7454	0.9690
Present	0.1408	0.3314	0.5029	0.6088	0.7575	0.9832

FG-A GPLR nanocomposite plates

	Mode Numbers (*m*, *n*)					
Source	(1, 1)	(2, 1)	(2, 2)	(3, 1)	(3, 2)	(3, 3)
Song et al. (2017a)	0.1118	0.2673	0.4110	0.5013	0.6299	0.8287
Present	0.1096	0.2624	0.4036	0.4926	0.6192	0.8154

The following dimensionless form of the natural frequency will be used in all of the illustrations:

$$\Omega = \omega \frac{a^2}{h} \sqrt{\frac{\rho_m}{E_m}} \qquad (16.1)$$

Figure 16.1 depicts the influence of the weight fraction of the GPLs on the natural frequency responses of FG-GPLR nanocomposite plates. The case of a square plate will be studied and the effect of changing the distribution pattern of the GPLs across the cross-sectional area of the plate on the dynamic response of the system is included. It can be found that increasing the content of the available GPLs in the media can result in an increase of the system's effective stiffness via a nonlinear slope. It is the remarkable stiffness of the nanosize GPLs that results in the

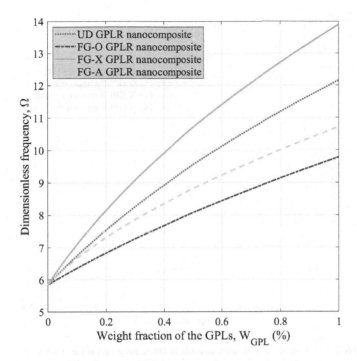

FIGURE 16.1 Variation of the dimensionless frequency of GPLR nanocomposite square plates versus weight fraction of the GPLs for various types of FG nanocomposite materials ($a/h = 10$, $\Delta T = 0$).

improvement of the total stiffness of the nanocomposite whenever a finite amount of them is added to any desired matrix. Hence, it is natural to observe an increase in the cross-sectional rigidities of the plate in this case and due to the direct relationship between the plate's rigidity and its natural frequency, the dynamic response of the system will be enhanced. As with our previous estimations, FG-X nanocomposite structures can support the greatest natural frequency followed by UD, FG-A, and FG-O ones, respectively.

Figure 16.2 shows the variation of the dimensionless frequency of the FG nanocomposite plates against temperature rise for the case of a square plate. Increasing the magnitude of the applied temperature rise will result in the reduction of the dimensionless natural frequency of the continuous system. This trend appears natural given the stiffness-softening impact of the temperature rise on the effective stiffness of the media. As can be seen in the figure, the natural frequency of the system will not dissipate entirely. The reason for this is that the addition of the GPLs strengthens the matrix such that the effects of adding the local temperature can be tolerated by the media and the structure will be able to support natural frequencies even when subjected to a 200 Kelvin temperature rise. As before, among all types of FG nanocomposite plates, FG-X is the most powerful one followed by UD, FG-A, and FG-O ones, respectively.

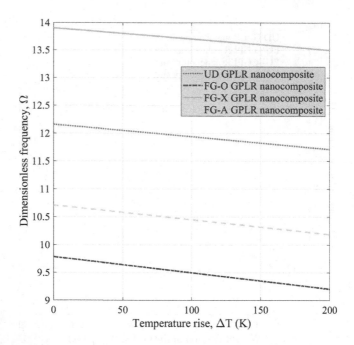

FIGURE 16.2 Variation of the dimensionless frequency of GPLR nanocomposite square plates versus temperature rise for various types of FG nanocomposite materials ($a/h = 10$, $W_{GPL} = 1\%$).

Figure 16.3 presents the effect of the length-to-thickness ratio of the plate on the dynamic responses of the nanocomposite structure once the effects of thermal loading are included. Realize that increasing the length-to-thickness ratio of the plate can result in an increase in the dimensionless natural frequency of the plate whenever no temperature gradient is available in the continuous system. This is because of the direct relation between the dimensionless frequency and the length-to-thickness ratio of the plate (refer to Eq. (16.1)). Note that this increasing trend will not be so clear once the length-to-thickness ratio of the plate exceeds approximately 20. On the other hand, it can be learned that increasing the temperature rise has a negative effect on the dynamic response of the nanocomposite structure. In other words, the natural frequency will be lessened once a nonzero value is assigned to the temperature rise and the reason for this trend is the softening influence of temperature on the total stiffness of the continuous system, which has a direct relation with the natural frequency of the plate. In cases where the magnitude of the temperature rise becomes bigger than a certain value, the thermal losses can result in complete dissipation of the system's fluctuations. It is clear that among all types of FG nanocomposite structures, the FG-X one can support higher natural frequencies followed by UD, FG-A, and FG-O ones, respectively.

As the final numerical example, the variation of dimensionless natural frequency of FG nanocomposite plates versus the aspect ratio of the structure is

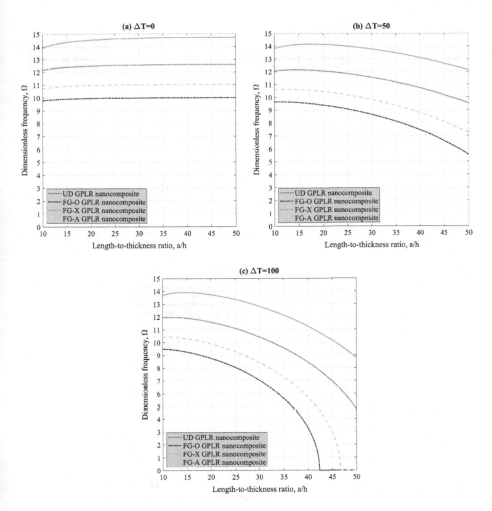

FIGURE 16.3 Variation of the dimensionless frequency of GPLR nanocomposite square plates versus length-to-thickness ratio for various types of FG nanocomposite materials (W_{GPL} = 1%).

plotted in Figure 16.4. According to this figure, increasing the aspect ratio of the plate will result in a reduction in the dynamic response of the system. This trend is not strange at all. Indeed, increasing the aspect ratio corresponds with lessening the influence of the longitudinal dimension of the plate on its natural frequency and this effect will result in a decrease in the dimensionless response of the system because of the direct relation between the length of the plate and its dimensionless frequency. Also, it can be observed that increasing the applied temperature rise will result in a decrease in the dynamic response of the system due to the softening impact of the thermal losses on the total stiffness of the nanocomposite plate. Furthermore, the order of the different types of nanocomposite materials in supporting the biggest natural frequency is as same as previous diagrams. In other words, the FG-X nanocomposite plates reveal the maximum frequency; whereas, FG-O ones support the minimum natural frequency.

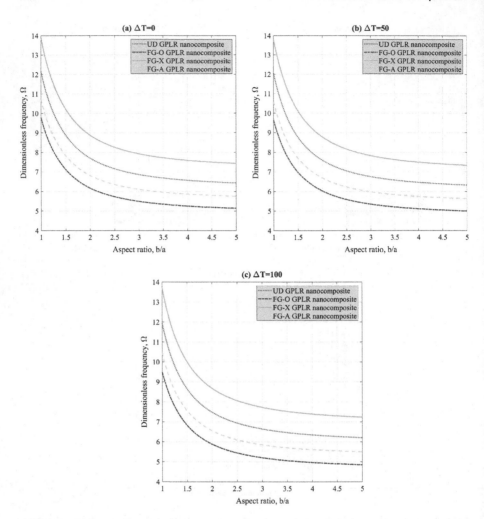

FIGURE 16.4 Variation of the dimensionless frequency of GPLR nanocomposite plates versus aspect ratio for various types of FG nanocomposite materials ($a/h = 10$, $W_{GPL} = 1\%$).

16.3 CONCLUDING REMARKS

In this chapter, a thermo-elastic vibration analysis was carried out on multilayered GPLR nanocomposite plates with FG distribution for the reinforcing elements. As in other chapters dealing with the GPLR nanocomposites, the effective material properties were derived based on the Halpin-Tsai micromechanical method for multilayered plate-type elements. The governing motion equations of the continuum were obtained by implementing the Reddy's plate hypothesis for the plates subjected to a thermal loading. In this chapter, the UTR-type thermal loading was considered. Afterward, the governing equations were solved via the Navier's solution to derive the natural frequency responses of simply supported nanocomposite plates. The most significant highlights of this chapter follow:

- The dimensionless natural frequency of the multilayered GPLR nanocomposite plate will increase when a greater amount of the GPLs is utilized when manufacturing the nanocomposite.
- Increasing the temperature rise results in a continuous decrease in the dynamic response of the system because of the softening effect of the temperature gradient on the total stiffness of the nanocomposite system.
- The dimensionless natural frequency of the plate can be reduced by increasing the aspect ratio of the plate. In this case, the plate will be more flexible and it is natural to see such a decreasing phenomenon.

17 Vibrational Behaviors of FG-GOR Nanocomposite Plates Embedded on an Elastic Substrate in Thermal Environment

17.1 PROBLEM DEFINITION

This chapter studies the dynamic responses of FG-GOR nanocomposite plates resting on a two-parameter elastic medium once the plate is subjected to a thermal gradient. The thermo-mechanical material properties of the plate are derived based on the instructions explained in Section 2.3.3 for the case of a single-layered nanocomposite sheet. The governing equations of the plate can be obtained using Hamilton's principle in association with the strain-displacement relations of refined shear deformable plate models. The impact of existence of thermal gradient on the natural frequency of the structure is considered, too. It is worth mentioning that three types of temperature rise are studied in the analysis. The set of the governing equations can be found by referring to Eqs. (2.215)–(2.218). After deriving the governing equations, the well-known Navier's type solution method will be utilized to calculate the natural frequency of the nanocomposite plates once all of the edges are simply supported. It will be shown that the dynamic response of the nanocomposite system can be dramatically influenced by varying the temperature rise.

17.2 NUMERICAL RESULTS AND DISCUSSION

In this section, some numerical examples will be presented for the purpose of understanding the mechanical behavior of the nanocomposite structure once the effects of various types of thermal loading are considered. The primary input of the problem (i.e., the material properties for both reinforcing GOs and host matrix) can be considered to be similar to those reported in Chapters 4 and 15. In this analysis, the total thickness of the plate-type element is $h = 2$ mm. For the sake of simplicity, the following dimensionless parameters will be implemented in the numerical illustrations:

$$\Omega = \omega \frac{L^2}{h} \sqrt{\frac{\rho_m}{E_m}}, \quad K_w = k_w \frac{L^4}{D^*}, \quad K_p = k_p \frac{L^2}{D^*}, \quad D^* = \frac{E_m h^3}{12(1 - v_m^2)} \quad (17.1)$$

TABLE 17.1

Comparison of the Dimensionless Fundamental Frequency of SSSS GPLR Nanocomposite Square Plates for Various Distribution Patterns of Nanofillers

Distribution Type	(García-Macías et al., 2018)	(Song et al., 2017a)	Present
Pure epoxy	0.058	0.058	0.057
UD	0.121	0.122	0.118
FG-O	0.097	0.102	0.100
FG-X	0.141	0.138	0.128
FG-A	0.117	0.112	0.118

First, the accuracy of the presented methodology will be shown. As can be seen in Table 17.1, the natural frequency values approximated by our model are in excellent agreement with those reported in the open literature. So, it can be guaranteed that this model is able to predict the natural frequency of the nanocomposite structures.

The first example show the variation of the dimensionless frequency versus weight fraction of the GOs once four environmental conditions are considered (see Figure 17.1). In these four cases, the effects of thermal gradient on the natural frequency are included as well as the effect of resting media on the dynamic response of the continua. According to this figure, the dimensionless frequency of the plate can be improved by increasing the content of the GOs available in the GOR nanocomposite material. However, the intensity of the dependency of the dynamic response on the weight fraction of the GOs is not identical for all types of GOR nanocomposites. Indeed, the biggest increase can be observed in the case of employing the FG-X pattern for the goal of reinforcing the matrix. After this type, the following order can be seen: UD GOR, FG-V GOR, and FG-O GOR nanocomposite, respectively. In addition, it can be seen that once the plate is subjected to a temperature rise, the dynamic response will decrease dramatically. The reason for this phenomenon is the increase of the system's equivalent compliance (i.e., identical with the reduction of the system's total stiffness) once the local temperature increases. On the other hand, it can be seen that adding an elastic foundation as the seat of the plate can help to establish stiffer plates that are able to support greater natural frequencies.

Figure 17.2 shows the variation of the dimensionless frequency of GOR nanocomposite plates against aspect ratio of the plate for various types of FG nanocomposites dismissing thermal losses. The plate is assumed to rest on an elastic foundation with dimensionless stiffnesses of $K_w = 50$ and $K_p = 5$. It is clear that the dimensionless frequency goes through a decreasing path as the aspect ratio of the plate becomes bigger. This phenomenon is logical due to the direct relation between the dimensionless form of the natural frequency and the plate's length, a. Indeed, the more is the aspect ratio, the more powerful will be the effect of the plate's width on the system's oscillations. Therefore, the reduction of the system's frequency is not strange at all. On the other hand, it can be seen that FG-X GOR nanocomposite plates will support

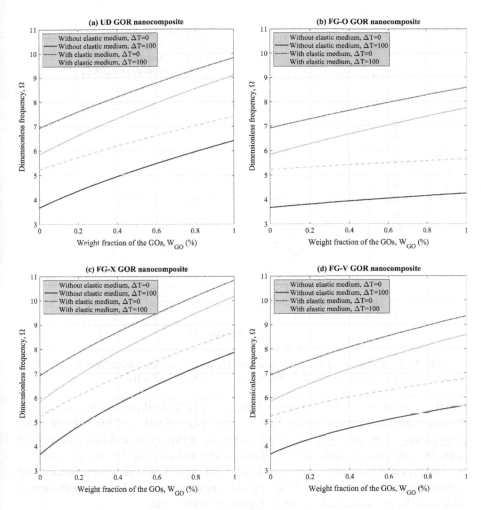

FIGURE 17.1 Variation of the dimensionless frequency of GOR nanocomposite square plates versus weight fraction of the GOs for various thermal environments and seats for the structure regarding the effects of various distribution patterns of FG nanocomposites ($a/h = 10$).

higher frequencies compared with other types of FG nanocomposites followed by UD GOR, FG-V GOR and FG-O GOR ones, respectively.

The next numerical example investigates the coupled influences of temperature rise and FG distribution patterns of GOs on the dimensionless frequency of the GOR nanocomposite square plates (see Figure 17.3). According to this figure, the worst thermal condition for the damping of the system's fluctuations occurs once the entire thickness of the plate is subjected to a uniform-type thermal gradient, also known as UTR. In this situation, the dimensionless frequency of the plate will be damped more easily than other types of thermal loading. After this type, LTR and STR will affect the dynamic response of the plate, respectively. In general, it can be observed

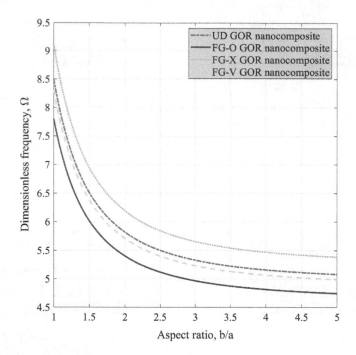

FIGURE 17.2 Variation of the dimensionless frequency versus aspect ratio for FG-GOR nanocomposite plates ($a/h = 10$, $K_w = 50$, $K_p = 5$, $W_{GO} = 0.5\%$, $\Delta T = 0$).

that the greatest oscillation frequency belongs to FG-X GOR nanocomposites in any desired temperature rise followed by UD GOR, FG-V GOR, and FG-O GOR ones, respectively. However, there exists an exception in this issue and this case can be seen once the plate is subjected to STR-type thermal loading. In fact, the order of last two types of FG nanocomposites, FG-V GOR and FG-O GOR, may be changed depending on the amplitude of the applied temperature rise. This phenomenon is related to the mathematical products appearing in this case.

The variation of the dimensionless frequency versus temperature rise for various weight fractions of the GOs is plotted in Figure 17.4 for different types of GOR nanocomposite plates. It is clear that increasing the weight fraction of the GOs results in stiffer plates that can support higher natural frequencies. Note that changing the weight fraction of the GOs can result in different effects on the frequency-temperature rise curve of FG-GOR nanocomposite structures. In other words, the final temperature that results in complete damping of the system's vibration cannot be tailored by tuning the weight fraction of the GOs and this parameter can only improve the value of the dimensionless frequency in smaller temperature rises. In the case of FG-X nanocomposite plates, the frequency-temperature rise curve of the plate will be shifted to the right-hand side and the critical temperature which results in dissipation of the system's vibration will be increased in addition to the magnitude of the natural frequency in any desired temperature rise. However, the behavior of the plates manufactured from FG-O and FG-V nanocomposites is a little different. Indeed, in these cases, there exists a thermal point where the trend of the diagram

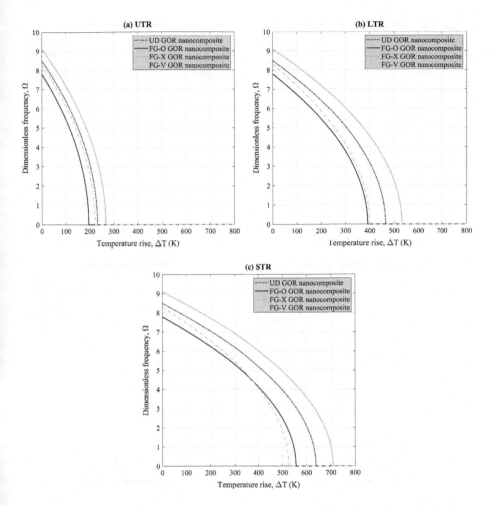

FIGURE 17.3 Variation of the dimensionless frequency versus temperature rise for various FG-GOR nanocomposite square plates ($a/h = 10$, $K_w = 50$, $K_p = 5$, $W_{GO} = 0.5\%$).

changes. Before this point, the vibration frequency of plates reinforced with a higher content of GOs is bigger; whereas in temperature rises after this point the order will be reversed. It can be concluded that in FG-O and FG-V nanocomposite structures, systems with greater amount of GOs possess higher frequency in the beginning and their frequency will be damped sooner at the end.

Finally, Figure 17.5 presents a quantitative study about the variation of the dimensionless frequency against temperature rise when the foundation parameters are varied. Based on this figure, the dimensionless frequency of the UD GOR nanocomposite plates can be enhanced once stiffer springs are employed for each of the Winkler or Pasternak terms. The main reason of this phenomenon is the stiffness improvement which will appear in the continuous system due to the enhancement of the system's structural stiffness. Obviously, the influence of the Pasternak-type spring is more than that of the Winkler-type one. In addition, it can

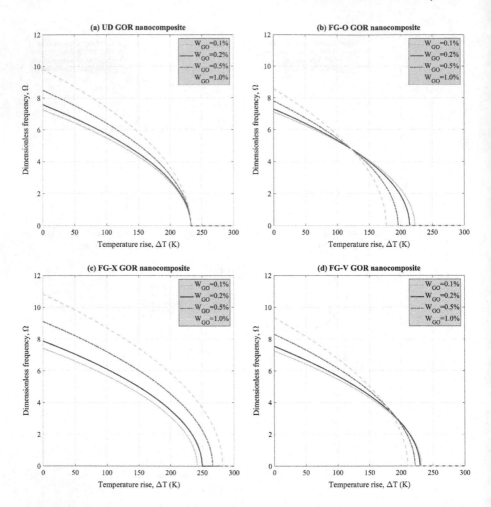

FIGURE 17.4 Variation of the dimensionless frequency versus uniform-type temperature rise for various weight fractions of the GOs regarding for different types of distribution patterns of square plates ($a/h = 10$, $K_w = 50$, $K_p = 5$).

be observed that the fluctuation of the plates subjected to a UTR will attenuate earlier than those subjected to LTR and STR, respectively. This issue can be seen in the previous diagrams, too.

17.3 CONCLUDING REMARKS

This chapter analyzed the thermal vibration of FG-GOR nanocomposite single-layered plates rested on a Winkler-Pasternak elastic seat. The equivalent material properties were derived based on the Halpin-Tsai micromechanical procedure. In addition, the extraction of the system's motion equations was carried out by combining the

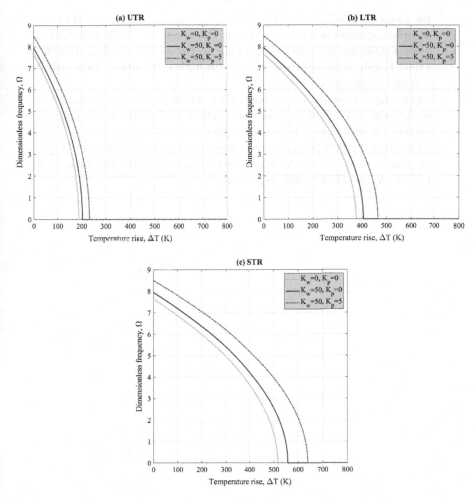

FIGURE 17.5 Variation of the dimensionless frequency of UD GOR nanocomposite plates versus temperature rise for various foundation parameters regarding for different types of thermal force applied on the structure ($a/h = 10$, $W_{GO} = 0.5\%$).

kinematic relations of refined higher-order plates with the concept of Hamilton's principle. Thereafter, the enriched eigenvalue problem was solved analytically for simply supported plates to reach the frequency of the vibrating system. A brief review of the most crucial highlights follows to emphasize the importance of the involved terms in estimating of the natural frequency of the system:

- The dimensionless frequency of the GOR nanocomposite plate will diminish once the amplitude of the temperature rise is increased.
- Increasing the content of the GOs in the composition of the constituent material results in an amplification of the dimensionless frequency for

the cases of FG-X and UD GOR nanocomposite structures. However, this phenomenon is only valid in the FG-O and FG-V GOR nanocomposites if the temperature rise does not exceed a critical value. After this value, the fluctuation of the plates with higher amount of GOs will be damped more easily.

- Increasing the foundation parameters will result in an amplification in the dimensionless frequency of the nanocomposite plate due to the enhancement of the structural stiffness of the plate.
- Dismissing thermal gradients, among various types of GOR nanocomposite plates, FG-X nanocomposites possess the peak response followed by UD, FG-V, and FG-O ones, respectively.

18 Coupled Impacts of Thermal and Viscose Losses on the Natural Frequency of FG-GOR Nanocomposite Plates

18.1 PROBLEM DEFINITION

This chapter analyzes the thermo-viscoelastic vibrational characteristics of FG-GOR nanocomposite plates. The plate is assumed to be constructed from FG nanocomposites and considered to be embedded on a visco-Pasternak medium. The equivalent material properties of the structure will be developed using the well-known Halpin-Tsai micromechanical method for a single-layered plate. Thereafter, the dynamic form of the principle of virtual work will be extended for a refined plate model to extract the set of the coupled governing equations of the plate's motion. Afterward, the Navier-type solution will be used to generate the natural frequencies of the nano-composite plate once all of the edges are assumed to be simply supported. The results of this chapter reveal a critical dependency of the system's frequency on the viscose coefficient of the viscoelastic substrate as well as thermal gradients.

18.2 NUMERICAL RESULTS AND DISCUSSION

This section investigates the influences of different terms on the behavior of the system's natural frequency. In the following numerical examples, the plate's total thickness is $h = 2$ mm. Also, the material properties are assumed to be as same as those utilized in the previous chapter. As with all of the previous chapters, the following dimensionless terms will be implemented in the future examples for the sake of simplicity:

$$\Omega = \omega \frac{a^2}{h} \sqrt{\frac{\rho_m}{E_m}}, \quad K_w = k_w \frac{a^4}{D^*}, \quad K_p = k_p \frac{a^2}{D^*}, \quad C_d = c_d \frac{a^2}{\sqrt{hD^* \rho_m}}, \quad D^* = \frac{E_m h^3}{12\left(1 - v_m^2\right)}$$

(18.1)

Figure 18.1 depicts the variation of the dimensionless frequency versus weight fraction of the GOs for a UD GOR nanocomposite square plate considering the impact of using various types of foundations on the mechanical response of the

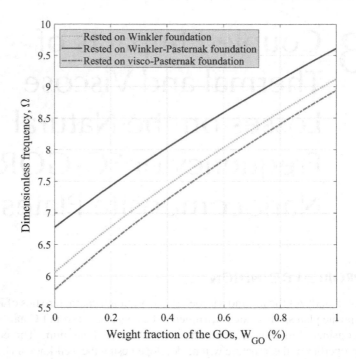

FIGURE 18.1 Variation of the dimensionless frequency of UD GOR nanocomposite square plates versus weight fraction of the GOs for various types of foundations ($a/h = 10$, $\Delta T = 10$).

continuous system. It can be observed that increasing the content of the GOs in the nanocomposite results in an amplification in the dynamic response of the system. The physical reason is the increasing role of the GOs in determining the equivalent stiffness of the system. It is not strange to see such a phenomenon due to the direct relationship between the equivalent stiffness of the system and its dynamic response. Moreover, plates rested on elastic foundation can reveal greater natural frequencies compared with those rested on a viscoelastic one. So, the diminishing influence of the viscose term of the visco-Pasternak can be clearly seen in this figure. In fact, once the nanocomposite plate rests on a visco-Pasternak substrate, the viscose losses caused by the damper will result in the attenuation of the system's oscillation.

The variation of dimensionless frequency versus damping coefficient is plotted in Figure 18.2 for various FG nanocomposite plates subjected to different types of thermal loading. According to this figure, the vibration of the system can be highly dissipated once the UTR is applied on the plate-type element followed by LTR and STR, respectively. In other words, the thermal losses can be better observed once all parts of the plate are subjected to a uniform-type temperature gradient. Once again, increasing the damping coefficient of the elastic medium can result in the dissipation of the natural frequency of the nanocomposite structure. It can be perceived that there exists a critical value for the damping coefficient such that after

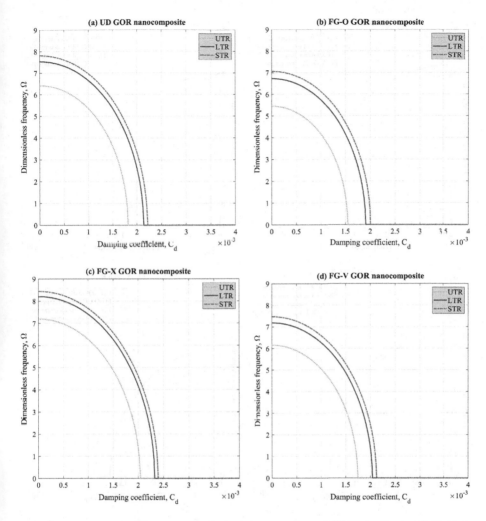

FIGURE 18.2 Variation of the dimensionless frequency of GOR nanocomposite square plates versus damping coefficient for various types of thermal gradient considering various distribution patterns of GOs across the thickness ($a/h = 10$, $W_{GO} = 0.5\%$, $K_w = 50$, $K_p = 5$).

that value the dimensionless frequency of the plate will be damped completely. This critical value is in its peak amount for FG-X nanocomposites followed by UD, FG-V, and FG-O ones, respectively. Also, among all types of nanocomposite plates, FG-X ones support greater frequencies followed by UD, FG-V, and FG-O ones, respectively.

In the next illustration (see Figure 18.3), the variation of dimensionless frequency of GOR nanocomposite plates against damping coefficient is plotted once various types of FG nanocomposite materials are employed and the structure is assumed to be subjected to a UTR with different amplitudes. It can be observed that the dynamic response of the system will be entirely damped once the damping coefficient of the

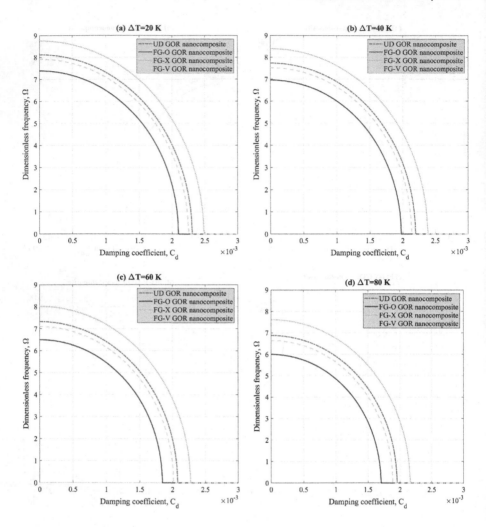

FIGURE 18.3 Variation of the dimensionless frequency of GOR nanocomposite square plates subjected to UTR versus damping coefficient considering various distribution patterns of GOs across the thickness ($a/h = 10$, $W_{GO} = 0.5\%$, $K_w = 50$, $K_p = 5$).

viscoelastic medium reaches its critical value. This value is not unique for each type of the FG nanocomposites and depends on the distribution pattern. The required damping coefficient for complete attenuation of the vibrations of FG-X GOR plates is greater than that of other types followed by UD GOR, FG-V GOR, and FG-O GOR plates, respectively. In addition, the dimensionless frequency of the system will be lessened when the amplitude of the applied temperature rise increases. This is the result of the softening influence of temperature on the material.

Figure 18.4 and Figure 18.5 show the quantitative impacts of Winkler and Pasternak coefficients on the dimensionless frequency of the UD GOR nanocomposite plates once the effects of both thermal and viscose losses are considered. It is obvious that increasing the stiffness of the foundation's springs can be one way

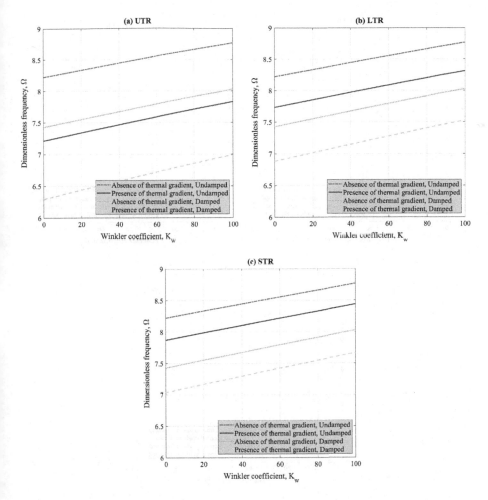

FIGURE 18.4 Variation of the dimensionless frequency of UD GOR nanocomposite square plates subjected to various types of temperature rise versus Winkler coefficient considering the influences of different types of visco-thermal losses on the dynamic response of the continuous system ($a/h = 10$, $W_{GO} = 0.5\%$, $K_p = 5$).

to amplify the vibration frequency due to the increase generated in the equivalent stiffness of the continuous system. In fact, the structural stiffness of the plate will be increased in the case of adding either Winkler or Pasternak coefficient. Also, it can be seen in both of these figures that increasing the temperature or the damping coefficient of the visco-Pasternak foundation can result in dissipating the dynamic response of the nanocomposite plate. In the case of applying a UTR, the effect of thermal losses is more powerful than that of the viscose one; however, this trend is not valid in the case of applying a LTR or STR on the nanocomposite system. In addition, it is clear that in an identical dimensionless coefficient, the Pasternak-type springs of the medium can improve the dynamic response of the system more than Winkler-type ones.

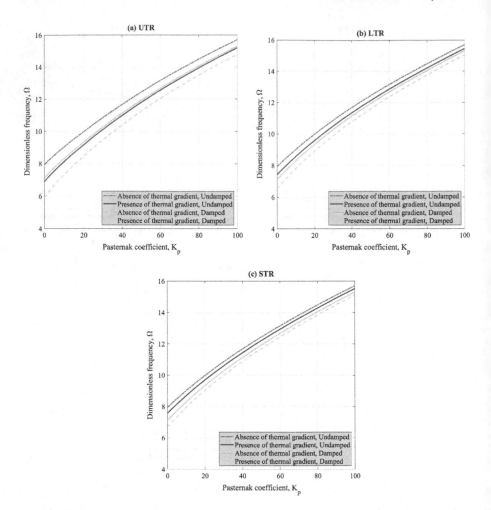

FIGURE 18.5 Variation of the dimensionless frequency of UD GOR nanocomposite square plates subjected to various types of temperature rise versus Pasternak coefficient considering the influences of different types of visco-thermal losses on the dynamic response of the continuous system ($a/h = 10$, $W_{GO} = 0.5\%$, $K_w = 100$).

18.3   CONCLUDING REMARKS

This chapter depicted the incorporated influences of viscose and thermal losses on the dynamic responses of FG-GOR nanocomposite plate-type elements. The homogenized thermo-elastic material properties of the FG nanocomposite material were derived based on the Halpin-Tsai micromechanical procedure for the case of a single-layered plate. Thereafter, the governing differential equations of the problem were extracted according to the expansion of Hamilton's principle for refined higher-order plates rested on a three-parameter viscoelastic medium. The impacts of various types of temperature gradient such as uniform, linear, and sinusoidal ones on the dynamic response of the problem were included, too. At the end, the governing equations were solved analytically to derive the natural frequency of the continuum. The highlights of the present chapter follow to emphasize the significance of the effects of the participating terms on the dynamic response of the system:

- Increasing the content of the existing nanosize GOs will result in amplification of the vibration frequency in any desired type of FG-GOR nanocomposite plate.
- If an engineer seeks to lessen the vibration frequency of the oscillating system by changing the local temperature, it is better to use the UTR rather than LTR and STR.
- An increase in the viscose coefficient of the viscoelastic substrate causes an attenuation in the natural frequency of the FG-GOR nanocomposite plate.
- Increasing each of the stiff coefficients of the foundation can intensify the dynamic response of the system. The effect of the Pasternak coefficient is greater than that of the Winkler one.
- Among different types of FG nanocomposites, the FG-X one can reveal higher natural frequencies followed by UD, FG-V, and FG-O, respectively.

19 Nonlinear Vibration Analysis of Porous FG-GOR Nanocomposite Shells

19.1 PROBLEM DEFINITION

The present chapter analyzes the nonlinear dynamic responses of cylindrical shells fabricated from porous GOR nanocomposite material. The porous nanocomposite will be considered to be reinforced via dispersion of GOs in three ways, namely non-uniform (type-I and type-II) and uniform (type-III). The equivalent material properties of the porous nanocomposite will be calculated using the instructions presented in Chapter 2, Section 2.3.5. The shell-type structure is assumed to be a thin-walled structure and the governing equations of the problem were derived based upon the classical shell theory. In this chapter, the effects of geometrical nonlinearity are included, implementing the von-Karman relations for the purpose of deriving the nonzero components of the strain tensor. The governing equations of the problem can be found in Eqs. (2.36)–(2.38). As expressed in Chapter 2, due to the nonlinearities of the governing equations, first they will be reduced to a unified equation and after that the final equation will be solved on the basis of the multiple scales method (i.e., one of the most famous perturbation-based numerical techniques). In the following sections, the effects of different parameters on the frequency ratio of the nanocomposite cylinders will be shown within the framework of a group of illustrations.

19.2 NUMERICAL RESULTS AND DISCUSSION

This section depicts the results of the nonlinear free vibration analysis in the framework of some numerical examples. In the following examples, the thickness of the shell-type element is assumed to be $h = 2$ mm. Also, the radius-to-thickness ratio and length-to-radius ratio will be $R/h = 100$ and $L/R = 20$, respectively; unless, another value is reported. The weight fraction of the GOs will be presumed to be equal with $W_{GO} = 1\%$. Before discussing the results of the present study, the validity of the results will be examined in a comparison study. Based on Table 19.1, the natural frequency responses of the present study are in a remarkable agreement with those reported in the literature. Hence, the presented investigations can be accepted as reliable results. Also, it can be observed that the smallest response corresponds with $m = 1$ and $n = 3$. Therefore, this mode will be considered in all of the following

TABLE 19.1

Comparison of the Linear Natural Frequencies (Hz) of Porous GPLR Nanocomposite Shells ($e_1 = 0.5$)

m	Source	\multicolumn{5}{c}{Circumferential Wave Number, n}
		1	2	3	4	5
1	Wang et al. (2019)	20.79	5.61	5.45	9.30	14.74
	Present	20.80	5.61	5.46	9.30	14.74
2	Wang et al. (2019)	77.55	20.87	10.60	10.66	15.15
	Present	77.56	20.88	10.61	10.67	15.16
3	Wang et al. (2019)	156.88	45.45	21.39	15.05	16.67
	Present	156.90	45.45	21.40	15.05	16.68

numerical examples to monitor the mechanical behavior of the fundamental natural frequency of the nanocomposite system.

In Figure 19.1, the influence of various types of porosity distributions is covered while drawing the variation of the frequency ratio versus dimensionless maximum deflection for different types of GOR nanocomposite shells. It can be inferred that the greatest mechanical response can be developed in the case of using GOR-B nanocomposite shells. In fact, the reinforcement goal can be better satisfied once this type of GOR nanocomposites is implemented. In other words, the stiffness enhancement is greater in this case among all types of GOR nanocomposites. On the other hand, in any desired GOR nanocomposite shell, the peak value of the frequency ratio corresponds with existence of type-II porosities in the media. The reason is that the destroying impact of the pores on the stiffness of the nanocomposite in this case is lower in comparison to other types of porous nanocomposites. Hence, it is preferred to have porous GOR-B nanocomposite shells with type-II distribution for the pores if omitting porosities in the media is impossible.

The variation of the frequency ratio versus dimensionless maximum deflection for different types of porous GOR-A nanocomposites is plotted in Figure 19.2. In this figure, the influence of changing the porosity coefficient, e_1, on the dynamic response of the system is highlighted. It can be seen that the highest frequency ratio can be attained with type-II distribution for the pores in the nanocomposite material. The physical reason can be observed in this figure. Indeed, the dynamic response of the system grows when a greater value is assigned to the porosity coefficient. This mathematical effect shows that given a GOR nanocomposite structure such that the big pores are at the top and bottom of the structure, the dynamic response of the system will increase when the porosity coefficient e_1 is increased. This phenomenon is the result of the direct relationship between coefficients e_1 and e_2. In other words, the porosity coefficient e_2 will increase as the coefficient e_1 is amplified. This relationship can be better observed by referring to the Eq. (2.275) or taking a brief look at Figure 2.1. It is natural to observe such an intensifying trend for the frequency ratio of type-II porous nanocomposites once the coefficient is increased since the increase

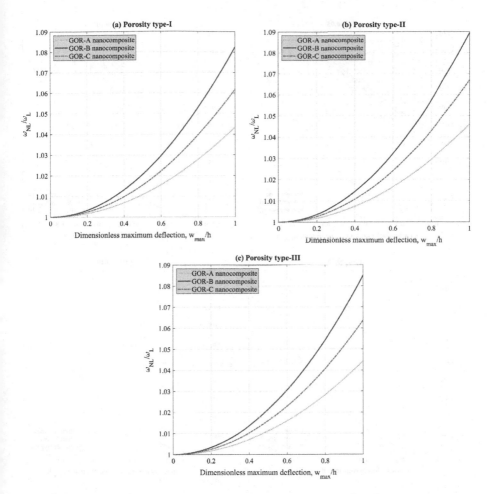

FIGURE 19.1 Variation of the fundamental frequency ratio of porous GOR nanocomposite shells versus dimensionless maximum deflection for various types of distribution patterns of GOs across the shell's thickness regarding for different types of porosities in the media ($m = 1$, $n = 3$, $R/h = 100$, $L/R = 20$, $e_1 = 0.1$, $W_{GO} = 1\%$).

of the porosity coefficient will decrease the effect of the porosity on the stiffness of type-II porous materials. In addition, due to the inverse relationship between e_3 and e_1, the frequency ratio will be intensified when the porosity coefficient is increased in type-III porous nanocomposite shells. The reason for this is that in this type of nanocomposite, the destructive effect of the porosities will be amplified when employing a great porosity coefficient, e_3. Due to the reduction of the porosity coefficient e_3 with increases of e_1, the dynamic response of GOR porous nanocomposites with type-III distribution for the pores will be enhanced as the coefficient e_1 increases. However, the amount of this increment is too small to be monitored easily.

The next numerical example investigates the effect of the porosity coefficient, e_1, on the variation of the frequency ratio versus dimensionless maximum deflection once

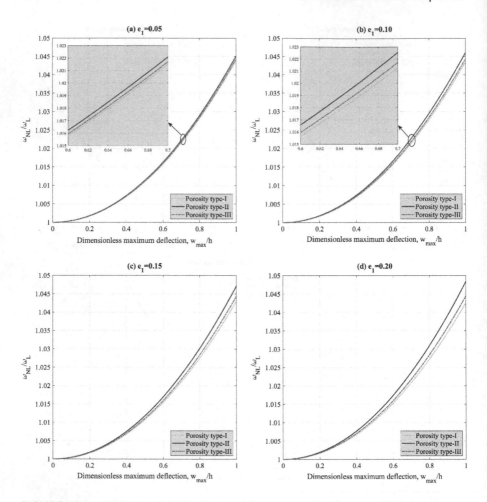

FIGURE 19.2 Variation of the fundamental frequency ratio of porous GOR-A nanocomposite shells versus dimensionless maximum deflection for various porosity types ($m = 1$, $n = 3$, $R/h = 100$, $L/R = 20$, $W_{GO} = 1\%$).

different types of porosity distribution are considered. It can be seen in Figure 19.3 that increasing the porosity coefficient will result in a reduction in the frequency ratio of the shell with type-I porosities because of the destructive impact of the existence of type-I pores in the continuum on the equivalent stiffness of the nanocomposite material. However, the dynamic response of the porous GOR nanocomposite shells will be enhanced whenever type-II and type-III porosity distributions are considered. The reason for this is that increasing the coefficient e_1 corresponds with increases of e_2 and the stiffness of the porous nanocomposite will less compromised once e_2 grows. On the other hand, increasing e_1 results in a reduction in the magnitude of e_3, which leads to lower destructive effects on the equivalent stiffness of the type-III porous GOR nanocomposites.

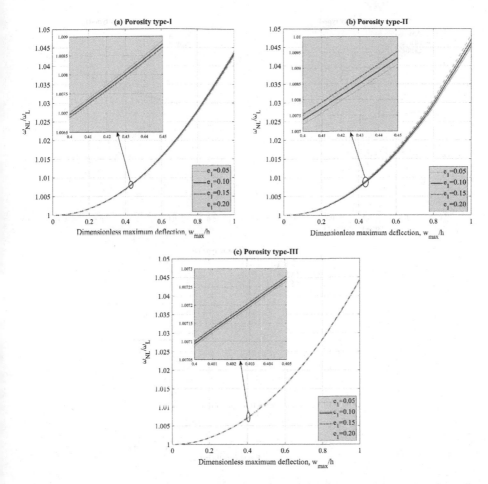

FIGURE 19.3 Variation of the fundamental frequency ratio of porous GOR-A nanocomposite shells versus dimensionless maximum deflection for various porosity coefficients ($m = 1$, $n = 3$, $R/h = 100$, $L/R = 20$, $W_{GO} = 1\%$).

Finally, the variation of the frequency ratio versus porosity coefficient e_1 for different types of porous GOR nanocomposite structures is plotted in Figure 19.4. Based on this figure, it can be seen that increasing the porosity coefficient results in a decreasing trend in the dynamic response of the nanocomposite system considering type-I distribution. This trend cannot be observed in type-II and type-III porous nanocomposites. In other words, the dynamic response of the continuous system in the aforementioned cases will be intensified once the porosity coefficient e_1 is added. The amount of this increase is clear in the case of type-II; while, it cannot be observed easily in type-III porous nanocomposite structures. The physical reasons of the above trends were previously discussed and it will not be explained anymore. Again, it can be understood that GOR-B nanocomposite systems can support the

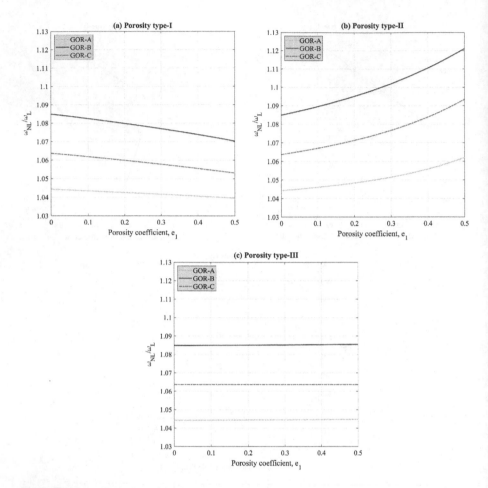

FIGURE 19.4 Variation of the fundamental frequency ratio of porous GOR nanocomposite shells versus porosity coefficient for various types of GOR nanocomposites ($m = 1$, $n = 3$, $R/h = 100$, $L/R = 20$, $W_{GO} = 1\%$).

greatest frequency ratio among all types of FG-GOR nanocomposite shells followed by GOR-C and GOR-A nanocomposite structures, respectively.

19.3 CONCLUDING REMARKS

In this chapter, the nonlinear vibration analysis of FG-GOR porous nanocomposite shells was accomplished in the framework of the classical shell theory for thin-walled shell-type elements. The equivalent material properties of the nanocomposite were obtained on the basis of the combination of the Halpin-Tsai micromechanical scheme with a porosity-based homogenization method. The set of governing differential equations of the problem were obtained by expanding the dynamic form of the principle of virtual work for classical theory of cylindrical shells regarding the impacts of the geometrical nonlinearities in the framework of implementing the von-Karman strains. Thereafter, the free vibration response of the problem was developed using the multiple scales method, as an efficient numerical method for the nonlinear dynamic problems. The following review will emphasize the crucial role of the participating variants in determining the natural frequency of the continuous system:

- The frequency ratio of the GOR-B nanocomposite structures is greater than that of GOR-C and GOR-A ones, respectively.
- The greatest dynamic response of the porous system can be obtained using type-II porosity distribution.
- Increasing the first porosity coefficient results in a gradual decrease of the dynamic response of the system whenever the type-I porous nanocomposite material is used for the fabrication. However, the dynamic response of the system can be intensified once type-II or type-III porosity distributions are considered.

20 Transient Analysis of Porous GOR Nanocomposite Plates

20.1 PROBLEM DEFINITION

This chapter investigates the dynamic bending behaviors of porous GOR nano-composite plate-type elements. The equivalent material properties of the porous GOR nanocomposite can be obtained using the instructions of Section 2.3.5 incorporated with Section 2.3.3. In this chapter, three types of porosity distributions are covered as well as three types of reinforcing for the employed nanocomposite structure. Also, the governing equations of the problem will be considered to be as same as Eqs. (2.215)–(2.218). In fact, the refined-type HSDT of the plates is utilized to derive the governing equations free from using any extra shear correction factor. Once the governing equations are produced, the transient response of the system can be derived using a two-step Laplace transformation technique, as discussed in Section 2.2.7. The abovementioned equations will be solved for the case of a plate with simply supported edges. In the numerical results it will be proven that the existence of pores in the nanocomposite system can result in higher deflections for the continuous system due to the destroying effects of porosity on the mechanical performance of the structure. The next section discusses the phenomenological trends that can be observed in the dynamic bending behaviors of the porous nanocomposite plate.

20.2 NUMERICAL RESULTS AND DISCUSSION

This section shows the influences of various types of GOR nanocomposite material and porosity distributions on the bending characteristics of plates manufactured from porous nanocomposite materials. The total thickness of the plate-type element is $h = 2$ mm and the length-to-thickness ratio is $a/h = 10$ unless another value is reported for these variants. To examine the validity of the presented results, the deflection responses of this modeling are compared with those reported by Park and Choi (2018). The results of the validation study are tabulated in Table 20.1. It can be seen that the results of this method are in a good agreement with those reported in the literature. The differences between the results originate from the fact that the present study is able to consider higher-order shear deformation; whereas, the research conducted by Park and Choi (2018) deals with a modified form of the FSDT. Hence, it is natural to see some tolerances while comparing the results.

225

TABLE 20.1

Comparison of the Dimensionless Central Deflection of SSSS Rectangular Plates

a/b	a/h	Reference	Central Deflection
0.5	5	Park and Choi (2018)	1.1430
		Present	1.1992
	10	Park and Choi (2018)	1.0454
		Present	1.0883
	25	Park and Choi (2018)	1.0181
		Present	1.0572
	1000	Park and Choi (2018)	1.0129
		Present	1.0512
1	5	Park and Choi (2018)	0.4904
		Present	0.5030
	10	Park and Choi (2018)	0.4273
		Present	0.4338
	25	Park and Choi (2018)	0.4096
		Present	0.4143
	1000	Park and Choi (2018)	0.4062
		Present	0.4106
2	5	Park and Choi (2018)	0.0958
		Present	0.1025
	10	Park and Choi (2018)	0.0714
		Present	0.0749
	25	Park and Choi (2018)	0.0646
		Present	0.0672
	1000	Park and Choi (2018)	0.0633
		Present	0.0657

In the following illustrations, the dimensionless form of the involved terms will be studied for the sake of simplicity:

$$W^*_{Dynamic} = \frac{E_m h w\left(\frac{a}{2},\frac{b}{2},t\right)}{1000 q_0 a^2}, \quad \tau = t\sqrt{\frac{E_m}{\rho_m a^2}} \qquad (20.1)$$

Figure 20.1 probes the variation of the dimensionless dynamic deflection against dimensionless time considering GOR-A porous nanocomposite plates. This diagram covers various types of porosity distributions and it is observed that the minimum deflection of the plate is shown when type-I porous nanocomposites are implemented. In fact, the destroying influence of the porosity on the equivalent stiffness of the media in this case is smaller than other cases. The maximum deflection amplitude corresponds with type-II porous nanocomposite plates. Generally, the deflection amplitude will increase when the porosity coefficient, e_1, is enlarged. This phenomenon shows the negative impact of the existence of porosities in the media on the deflection behaviors of the continuous system. It is clear that because there exists

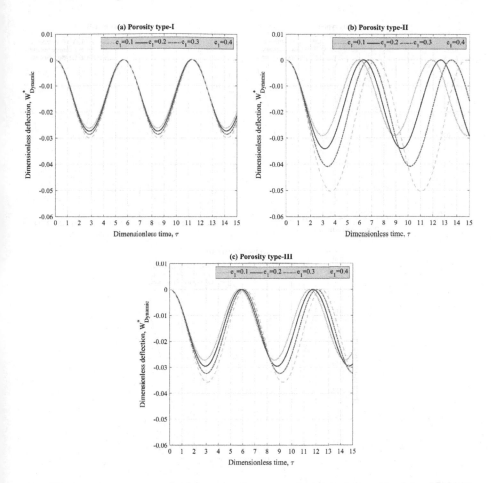

FIGURE 20.1 Variation of the dynamic deflection of GOR-A nanocomposite square plates versus dimensionless time for various porosity coefficients and various types of porous nanocomposites ($a/h = 10$, $W_{GO} = 1\%$).

no damping parameter in the governing equations, the oscillation of the system will not diminish and the system will keep on fluctuating as time proceeds.

The variation of the dynamic bending deflection versus time for various types of GOR porous nanocomposite plates is drawn in Figure 20.2. Based upon this diagram, it can be seen that GOR-B nanocomposite plates experience the maximum deflection among all types of GOR nanocomposite plates. In addition, type-II porous nanocomposite plates will be affected by the existence of porosities more than other types of porous nanocomposite systems. Hence, the possibility of having GOR-B porous nanocomposite plates with type-II distribution for the pores must be considered by the designers of dynamic devices like plate-type elements.

In the next example, the combined impacts of the porosity coefficient and type of porosity distribution on the dynamic deflection characteristics of the plate is observed by plotting the variation of the dimensionless dynamic deflection versus dimensionless time for various types of porosity distributions when the porosity coefficient

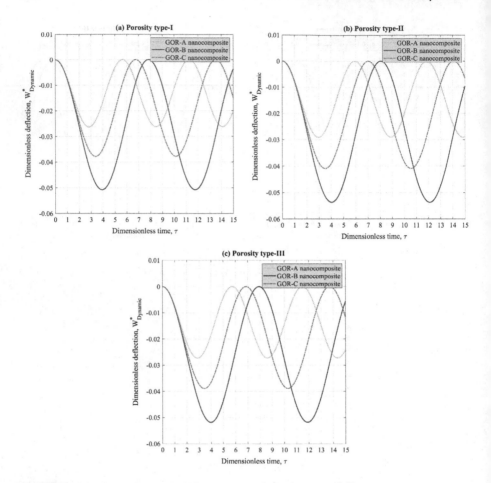

FIGURE 20.2   Variation of the dynamic deflection of GOR nanocomposite square plates versus dimensionless time for various types of GOR nanocomposites considering different porosity distributions ($a/h = 10$, $W_{GO} = 1\%$, $e_1 = 0.1$).

is varied (see Figure 20.3). According to this figure, the destroying effect of the presence of pores in the media can be well observed. In other words, the deflection amplitude will grow as a higher value is assigned to the porosity coefficient. Meanwhile, the worst bending condition corresponds with GOR-A nanocomposites with pores distributed in a type-II distribution. So, to consider the most critical circumstance, porous type-II nanocomposite plates must be considered when designing a fluctuating plate-type element manufactured from GOR nanocomposites.

At the end of this section, Figure 20.4 shows the effects of the geometry of the plate and the presence or absence of pores in the nanocomposite type-A material on the deflection characteristics of plates fabricated from such materials. It is obvious that the existence of porosities in the media will result in an increase in the deflection amplitude at any desired time. The effect of the aspect ratio of the nanocomposite plate can be well observed in this figure. In fact, the greater the width-to-length

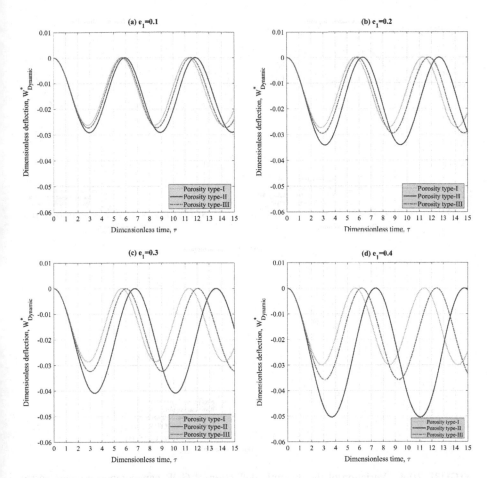

FIGURE 20.3 Variation of the dynamic deflection of GOR-A nanocomposite square plates versus dimensionless time for various types of porosity distributions and different porosity coefficients ($a/h = 10$, $W_{GO} = 1\%$).

ratio of the plate, the higher the deflection of the system. The reason is that rectangular plates are more flexible than square ones and because of the inverse relation between flexibility (i.e., identical with compliance) and stiffness, the system will lose its structural stiffness once the ratio of width-to-length becomes bigger and experiences higher deflections consequently.

20.3 CONCLUDING REMARKS

This chapter was concerned with the dynamic bending analysis of porous GOR nanocomposite plates. The material properties of the system were developed by implementing the Halpin-Tsai micromechanical theorem in association with the porosity-based homogenization technique. Three types of porous nanocomposite structures with graded material properties were covered in this chapter. Also, the

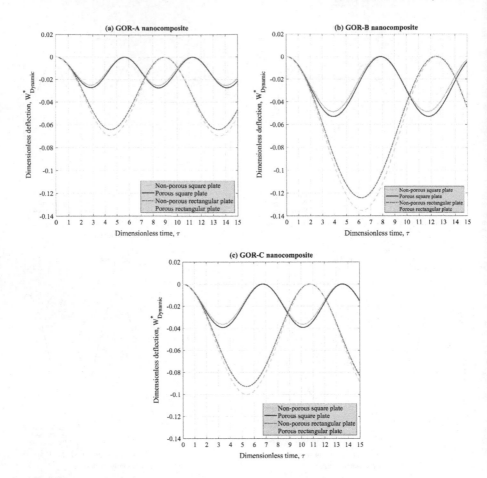

FIGURE 20.4 Variation of the dynamic deflection of GOR nanocomposite plates versus dimensionless time for various types of GOR nanocomposites and different aspect ratios for the structure ($a/h = 10$, $W_{GO} = 1\%$).

motion equations of the plate were assumed to be derived on the basis of the refined HSDT of the plates for the purpose of including the shear deflection effects while investigating the results. The most crucial highlights of this chapter follow:

- The porous nanocomposite plates will experience greater dynamic deflections due to the destroying effect of the pores on the stiffness of the media.
- The type-II porosity distribution will affect the deflection amplitude of the plate more than other types of porous nanocomposites.
- GOR-B nanocomposite plates will experience higher deflections compared to other types of GOR polymeric nanocomposite systems.
- Increasing the porosity coefficient results in a reduction in the stiffness of the continuum and enlarges the deflection amplitude.
- Rectangular non-square nanocomposite plates have higher deflections compared to square ones.

21 On Transient Responses of Viscoelastic GPLR Nanocomposite Plates

21.1 PROBLEM DEFINITION

This chapter presents an analytical solution for the damped dynamic bending behaviors of GPLR nanocomposite rectangular plates. The damping effects are generated due to the internal damping of the polymeric nanocomposite and these effects are covered in this chapter by implementing a viscoelastic model for the linear elastic nanocomposite materials. The effective material properties of the nanocomposite in the elastic domain can be easily generated using the fundamentals of the Halpin-Tsai micromechanical method for the case of a single-layered GPLR nanocomposite (for more information, refer to Section 2.3.2). The refined-type HSDT of the plates will be utilized to derive the governing equations of the plate, as expressed in Eqs. (2.222)–(2.225). However, note that the aforementioned set of governing equations belongs to the dynamic problem of an elastic GPLR nanocomposite and these equations must be changed to reach the governing equations for the damped bending analysis of a viscoelastic GPLR nanocomposite plate. This transfer can be simply accomplished using the instructions presented in Section 2.3.6 for viscoelastic nanocomposite materials. Based on the referenced section, the cross-sectional rigidities of the plate must be multiplied by the term $\left(1 + g\frac{\partial}{\partial t}\right)$ to derive the governing equations of a viscoelastic nanocomposite plate. In addition, the influence of the applied transverse bending force must be considered as mentioned at the end of Section 2.2.7 for the transient analysis of a refined plate. Following the previous steps, the final governing equations will be obtained and they will be solved analytically by combining the Navier-type solution with the concept of the Laplace transformation to produce the time-dependent deflection of the viscoelastic nanocomposite plate. In the following section, a group of numerical examples will clearly depict the effects of the material's internal damping on the dynamic responses of the nanocomposite plate.

21.2 NUMERICAL RESULTS AND DISCUSSION

In this section, the effects of various terms on the transient responses of the plate, in particular the internal damping coefficient of the viscoelastic nanocomposite, will be presented via a group of illustrations. The material properties of the GPLR nanocomposite are the same as those reported in the previous chapters concerned with the mechanical performance of GPLR nanocomposite continuous systems.

Also, the entire thickness of the structure is $h = 2$ mm and it will remain constant in all of the illustrations. In all of the case studies, the following dimensionless terms are defined:

$$W^*_{Dynamic} = \frac{E_m hw\left(\frac{a}{2}, \frac{b}{2}, t\right)}{1000 q_0 a^2}, \quad \tau = t\sqrt{\frac{E_m}{\rho_m a^2}}, \quad g = Ga^2\sqrt{\frac{\rho_m h}{D^*}}, \quad D^* = \frac{E_m h^3}{12\left(1 - v_m^2\right)} \quad (21.1)$$

where G is the structural damping of the material.

As the first example, Figure 21.1 shows the variation of the dynamic deflection of the GPLR nanocomposite square plates against time for various weight fractions of the GPLs once the influences of the material's internal damping are included. It can be seen that increasing the content of the GPLs in the composition of the polymeric nanocomposite will result in a reduction in the deflection amplitude of the viscoelastic plate. The physical interpretation of this trend is the enhancement that is created in the stiffness of the nanocomposite material by increasing the content of the GPLs. The plate will behave more stiffly and its resistance against bending will be bigger. Consequently, it is natural to see a decrease in the deflection amplitude of the nanocomposite plate in the case of increasing the GPLs. Note that the deflection of the nanocomposite structure will reach a stationary amplitude as time proceeds. This is due to the damping impact of the viscoelastic nature of the nanocomposite on

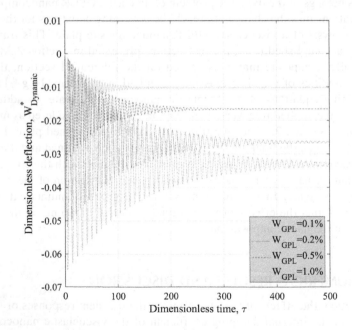

FIGURE 21.1 Variation of the dynamic deflection of viscoelastic GPLR nanocomposite square plates versus dimensionless time for various weight fractions of the GPLs in the media ($a/h = 10$, $G = 0.001$).

the dynamic response of the system. According to the figure, attaining the nanocomposite plate's stationary deflection will take less time as the content of the available GPLs in the nanocomposite grows.

Figure 21.2 illustrates the variation of the dynamic deflection of the viscoelastic GPLR nanocomposite plate versus time for different amounts of the structural damping. In this figure, three types of viscoelastic GPLR nanocomposite material are implemented whose difference is in the amount of the GPLs used in the fabrication of the nanocomposite. As understood from the previous figure, once more it can be certified that increasing the amount of the GPLs in the composition of the material can result in a more controlled fluctuation for the transient response of the system. The reason for this was explained before and it will not be discussed further. It can be seen that the dynamic deflection of the system will reach its stationary state sooner as the structural damping coefficient of the nanocomposite is increased. In other words, the amplitude of the system's

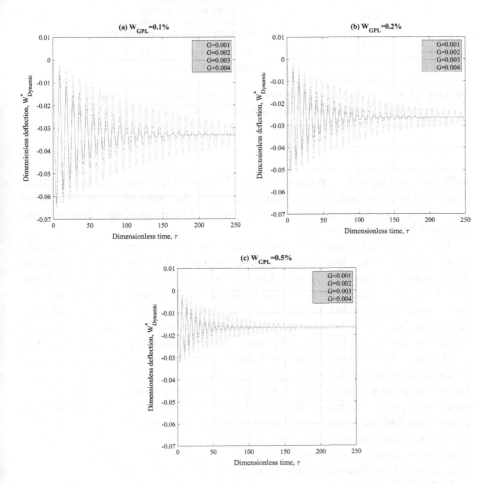

FIGURE 21.2 Variation of the dynamic deflection of viscoelastic GPLR nanocomposite square plates versus dimensionless time for various structural damping coefficients ($a/h = 10$).

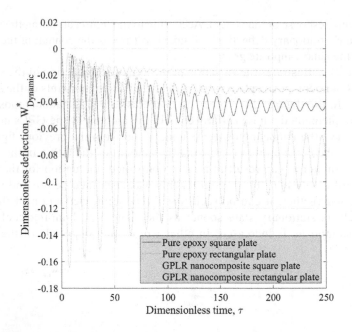

FIGURE 21.3 Variation of the dynamic deflection of viscoelastic GPLR nanocomposite plates versus dimensionless time for various types of pure and nanocomposite plates ($a/h = 10$, $G = 0.002$).

deflection will be more constrained when a greater value is assigned to the structural damping coefficient of the material. Again, the interpretations of the other phenomena depicted in this figure were discussed in the previous paragraph and they will be left for the readers.

The next example investigates the effect of the material's type and geometrical characteristics of the plate on the dynamic deflection of the viscoelastic nanocomposite plates (see Figure 21.3). Based on this figure, it is clear that GPLR nanocomposite plates are better choices whenever the dynamic deflection of the plate-type device must be controlled. In such cases, the plate will experience lower deflection amplitudes because of the stiffness improvement generated in the continuous system by adding nanosize GPLs to the polymeric matrix. On the other hand, the deflection amplitude of the rectangular plates is greater than that of the square ones. This is the result of the change in the flexibility of the continuum once its aspect ratio is varied. In fact, the plate will be more flexible whenever the ratio between the width and the length of it grows. Moreover, the flexibility possesses a direct relationship with the compliance and due to the inverse relation between the compliance and stiffness in the linear solids, the stiffness of the rectangular plates is smaller than that of the square ones. Hence, it is completely natural to see that rectangular plates will experience greater deflection amplitudes in comparison with square ones.

In the final example, the variation of the dimensionless dynamic deflection of the GPLR nanocomposite square plates versus dimensionless time is drawn in

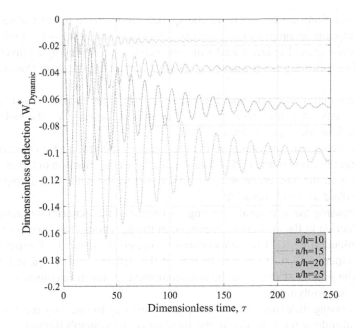

FIGURE 21.4 Variation of the dynamic deflection of viscoelastic GPLR nanocomposite square plates versus dimensionless time for various length-to-thickness ratios ($G = 0.002$, $W_{GPL} = 0.5\%$).

Figure 21.4 when the length-to-thickness ratio of the plate is varied. According to this diagram, the dimensionless dynamic deflection of the plate will increase as the length-to-thickness ratio of the plate is increased. The reason can be simply understood by referring to the definition of the dimensionless deflection, introduced in Eq. (21.1). Increasing the length-to-thickness ratio of the plate will result in an increase of the time required to see the stationary deflection of the plate once the effects of internal damping are included. It may be interesting to point out that the effect of changing the length-to-thickness ratio is not a linear effect. In other words, the amount of the increase of the system's deflection amplitude in the case of changing the length-to-thickness ratio from 10 to 15 is not identical with the situation in which the length-to-thickness ratio is changed from 20 to 25.

21.3 CONCLUDING REMARKS

This chapter investigated the viscoelastically influenced dynamic bending characteristics of GPLR nanocomposite plates. In this chapter, the Halpin-Tsai method was employed to derive the material properties of GPLR nanocomposite single-layered materials. Afterward, the governing equations of the transient problem of a plate were derived using a refined plate hypothesis in association with the concept of Hamilton's principle for a viscoelastic nanocomposite plate subjected to a distributed transverse bending force. The problem was solved within the framework of a two-step solution; the first step for the spatial decomposition and the second step for the temporal decomposition. The time-dependent responses of the problem were developed using the Laplace transformation technique followed by an inversion procedure. The following list presents the most important highlights of this chapter:

- The dynamic deflection of the plate will reach a stationary value as time proceeds once the influences of the internal damping of the polymeric GPLR nanocomposite are included.
- The deflection amplitude of the plate can be controlled by changing the weight fraction of the GPLs. The greater the weight fraction of the available GPLs in the nanocomposite, the lower will be the deflection amplitude of the plate under observation.
- Increasing the structural damping coefficient of the material will result in a reduction in the deflection amplitude of the nanocomposite structure.
- Implementing GPLR nanocomposite materials instead of pure epoxy polymers will result in a reduction of the deflection of the plate because of the greater stiffness of the nanocomposite material in comparison with the pure polymer.
- Increasing the aspect ratio of the plate is one way to intensify the deflection amplitude of the plate due to the increase of the system's flexibility.
- Similar to the aspect ratio, the deflection amplitude of the plate will grow dramatically when the length-to-thickness ratio of the plate is increased.

References

Ahmadi-Moghadam B, Sharafimasooleh M, Shadlou S, et al. (2015) Effect of functionalization of graphene nanoplatelets on the mechanical response of graphene/epoxy composites. *Materials & Design (1980-2015)* 66: 142–149.

Ahmadi-Moghadam B and Taheri F. (2015) Influence of graphene nanoplatelets on modes I, II and III interlaminar fracture toughness of fiber-reinforced polymer composites. *Engineering Fracture Mechanics* 143: 97–107.

Ajayan PM, Schadler LS, Giannaris C, et al. (2000) Single-walled carbon nanotube–polymer composites: Strength and weakness. *Advanced Materials* 12: 750–753.

Alesadi A, Galehdari M and Shojaee S. (2017) Free vibration and buckling analysis of cross-ply laminated composite plates using Carrera's unified formulation based on isogeometric approach. *Computers & Structures* 183: 38–47.

Alibeigloo A and Liew KM. (2013) Thermoelastic analysis of functionally graded carbon nanotube-reinforced composite plate using theory of elasticity. *Composite Structures* 106: 873–881.

Ansari R, Faghih Shojaei M, Mohammadi V, et al. (2014) Nonlinear forced vibration analysis of functionally graded carbon nanotube-reinforced composite Timoshenko beams. *Composite Structures* 113: 316–327.

Ansari R, Hasrati E, Faghih Shojaei M, et al. (2015) Forced vibration analysis of functionally graded carbon nanotube-reinforced composite plates using a numerical strategy. *Physica E: Low-dimensional Systems and Nanostructures* 69: 294–305.

Ansari R, Pourashraf T, Gholami R, et al. (2016) Analytical solution for nonlinear postbuckling of functionally graded carbon nanotube-reinforced composite shells with piezoelectric layers. *Composites Part B: Engineering* 90: 267–277.

Ansari R and Torabi J. (2016) Numerical study on the buckling and vibration of functionally graded carbon nanotube-reinforced composite conical shells under axial loading. *Composites Part B: Engineering* 95: 196–208.

Arash B, Wang Q and Varadan VK. (2014) Mechanical properties of carbon nanotube/polymer composites. *Scientific Reports* 4: 6479.

Aref AJ and Alampalli S. (2001) Vibration characteristics of a fiber-reinforced polymer bridge superstructure. *Composite Structures* 52: 467–474.

Aydogdu M. (2007) Thermal buckling analysis of cross-ply laminated composite beams with general boundary conditions. *Composites Science and Technology* 67: 1096–1104.

Aydogdu M. (2009) A new shear deformation theory for laminated composite plates. *Composite Structures* 89: 94–101.

Bakshi SR and Agarwal A. (2011) An analysis of the factors affecting strengthening in carbon nanotube reinforced aluminum composites. *Carbon* 49: 533–544.

Balzani C and Wagner W. (2008) An interface element for the simulation of delamination in unidirectional fiber-reinforced composite laminates. *Engineering Fracture Mechanics* 75: 2597–2615.

Barai P and Weng GJ. (2011) A theory of plasticity for carbon nanotube reinforced composites. *International Journal of Plasticity* 27: 539–559.

Barati MR and Zenkour AM. (2017) Post-buckling analysis of refined shear deformable graphene platelet reinforced beams with porosities and geometrical imperfection. *Composite Structures* 181: 194–202.

Beardmore P. (1986) Composite structures for automobiles. *Composite Structures* 5: 163–176.

Bisht A, Srivastava M, Kumar RM, et al. (2017) Strengthening mechanism in graphene nano-platelets reinforced aluminum composite fabricated through spark plasma sintering. *Materials Science and Engineering: A* 695: 20–28.

Brinson HF and Brinson LC. (2015) *Polymer engineering science and viscoelasticity*: Springer.

Ćetković M and Vuksanović D. (2009) Bending, free vibrations and buckling of laminated composite and sandwich plates using a layerwise displacement model. *Composite Structures* 88: 219–227.

Chai GB and Yap CW. (2008) Coupling effects in bending, buckling and free vibration of generally laminated composite beams. *Composites Science and Technology* 68: 1664–1670.

Chandrasekaran S, Sato N, Tölle F, et al. (2014) Fracture toughness and failure mechanism of graphene based epoxy composites. *Composites Science and Technology* 97: 90–99.

Chatterjee S, Nafezarefi F, Tai NH, et al. (2012a) Size and synergy effects of nanofiller hybrids including graphene nanoplatelets and carbon nanotubes in mechanical proper-ties of epoxy composites. *Carbon* 50: 5380–5386.

Chatterjee S, Wang JW, Kuo WS, et al. (2012b) Mechanical reinforcement and thermal con-ductivity in expanded graphene nanoplatelets reinforced epoxy composites. *Chemical Physics Letters* 531: 6–10.

Dano ML and Hyer MW. (2002) Snap-through of unsymmetric fiber-reinforced composite laminates. *International Journal of Solids and Structures* 39: 175–198.

Dano ML and Hyer MW. (2003) SMA-induced snap-through of unsymmetric fiber-reinforced composite laminates. *International Journal of Solids and Structures* 40: 5949–5972.

Daoush WM, Lim BK, Mo CB, et al. (2009) Electrical and mechanical properties of carbon nanotube reinforced copper nanocomposites fabricated by electroless deposition pro-cess. *Materials Science and Engineering: A* 513-514: 247–253.

Ebrahimi F, Dabbagh A and Civalek Ö. (2019a) Vibration analysis of magnetically affected graphene oxide-reinforced nanocomposite beams. *Journal of Vibration and Control* 25(23–24), 2837–2849. doi :10.1177/1077546319861002.

Ebrahimi F, Nouraei M and Dabbagh A. (2019b) Modeling vibration behavior of embed-ded graphene-oxide powder-reinforced nanocomposite plates in thermal environment. *Mechanics Based Design of Structures and Machines*: 1–24.

Ebrahimi F, Nouraei M and Dabbagh A. (2019c) Thermal vibration analysis of embedded graphene oxide powder-reinforced nanocomposite plates. *Engineering with Computers*.

Ebrahimi F, Nouraei M, Dabbagh A, et al. (2019d) Thermal buckling analysis of embedded graphene-oxide powder-reinforced nanocomposite plates. *Advances in Nano Research* 7: 293–310.

Ebrahimi F and Qaderi S. (2019) Stability analysis of embedded graphene platelets rein-forced composite plates in thermal environment. *The European Physical Journal Plus* 134: 349.

Fantuzzi N, Tornabene F, Bacciocchi M, et al. (2017) Free vibration analysis of arbitrarily shaped functionally graded carbon nanotube-reinforced plates. *Composites Part B: Engineering* 115: 384–408.

Fazzolari FA and Carrera E. (2011) Advanced variable kinematics Ritz and Galerkin formu-lations for accurate buckling and vibration analysis of anisotropic laminated composite plates. *Composite Structures* 94: 50–67.

Feng C, Kitipornchai S and Yang J. (2017a) Nonlinear bending of polymer nanocompos-ite beams reinforced with non-uniformly distributed graphene platelets (GPLs). *Composites Part B: Engineering* 110: 132–140.

Feng C, Kitipornchai S and Yang J. (2017b) Nonlinear free vibration of functionally graded polymer composite beams reinforced with graphene nanoplatelets (GPLs). *Engineering Structures* 140: 110–119.

Ferreira AJM, Roque CMC and Jorge RMN. (2005) Free vibration analysis of symmetric laminated composite plates by FSDT and radial basis functions. *Computer Methods in Applied Mechanics and Engineering* 194: 4265–4278.

Formica G, Lacarbonara W and Alessi R. (2010) Vibrations of carbon nanotube-reinforced composites. *Journal of Sound and Vibration* 329: 1875–1889.

Garcés JM, Moll DJ, Bicerano J, et al. (2000) polymeric nanocomposites for automotive applications. *Advanced Materials* 12: 1835–1839.

García-Macías E, Rodríguez-Tembleque L and Sáez A. (2018) Bending and free vibration analysis of functionally graded graphene vs. carbon nanotube reinforced composite plates. *Composite Structures* 186: 123–138.

Gholami R and Ansari R. (2017) Large deflection geometrically nonlinear analysis of functionally graded multilayer graphene platelet-reinforced polymer composite rectangular plates. *Composite Structures* 180: 760–771.

Gholami R and Ansari R. (2018) Nonlinear harmonically excited vibration of third-order shear deformable functionally graded graphene platelet-reinforced composite rectangular plates. *Engineering Structures* 156: 197–209.

Goh CS, Wei J, Lee LC, et al. (2005) Development of novel carbon nanotube reinforced magnesium nanocomposites using the powder metallurgy technique. *Nanotechnology* 17: 7–12.

Gojny FH, Wichmann MHG, Köpke U, et al. (2004) Carbon nanotube-reinforced epoxy-composites: enhanced stiffness and fracture toughness at low nanotube content. *Composites Science and Technology* 64: 2363–2371.

Guo H, Cao S, Yang T, et al. (2018) Vibration of laminated composite quadrilateral plates reinforced with graphene nanoplatelets using the element-free IMLS-Ritz method. *International Journal of Mechanical Sciences* 142–143: 610–621.

Hadden CM, Klimek-McDonald DR, Pineda EJ, et al. (2015) Mechanical properties of graphene nanoplatelet/carbon fiber/epoxy hybrid composites: Multiscale modeling and experiments. *Carbon* 95: 100–112.

Han Y and Elliott J. (2007) Molecular dynamics simulations of the elastic properties of polymer/carbon nanotube composites. *Computational Materials Science* 39: 315–323.

Haque A and Ramasetty A. (2005) Theoretical study of stress transfer in carbon nanotube reinforced polymer matrix composites. *Composite Structures* 71: 68–77.

Heydarpour Y, Mohammadi Aghdam M and Malekzadeh P. (2014) Free vibration analysis of rotating functionally graded carbon nanotube-reinforced composite truncated conical shells. *Composite Structures* 117: 187–200.

Jam JE and Kiani Y. (2015a) Buckling of pressurized functionally graded carbon nanotube reinforced conical shells. *Composite Structures* 125: 586–595.

Jam JE and Kiani Y. (2015b) Low velocity impact response of functionally graded carbon nanotube reinforced composite beams in thermal environment. *Composite Structures* 132: 35–43.

Joshi M and Chatterjee U. (2016) 8 - Polymer nanocomposite: An advanced material for aerospace applications. In: Rana S and Fangueiro R (eds) *Advanced Composite Materials for Aerospace Engineering*. Woodhead Publishing, 241–264.

Kamar NT, Hossain MM, Khomenko A, et al. (2015) Interlaminar reinforcement of glass fiber/epoxy composites with graphene nanoplatelets. *Composites Part A: Applied Science and Manufacturing* 70: 82–92.

Kausar A, Rafique I and Muhammad B. (2017) Aerospace application of polymer nanocomposite with carbon nanotube, graphite, graphene oxide, and nanoclay. *Polymer-Plastics Technology and Engineering* 56: 1438–1456.

Ke L-L, Yang J and Kitipornchai S. (2010) Nonlinear free vibration of functionally graded carbon nanotube-reinforced composite beams. *Composite Structures* 92: 676–683.

Ke L-L, Yang J and Kitipornchai S. (2013) Dynamic stability of functionally graded carbon nanotube-reinforced composite beams. *Mechanics of Advanced Materials and Structures* 20: 28–37.

Kim KT, Cha SI, Hong SH, et al. (2006) Microstructures and tensile behavior of carbon nanotube reinforced Cu matrix nanocomposites. *Materials Science and Engineering: A* 430: 27–33.

Kim M, Park Y-B, Okoli OI, et al. (2009) Processing, characterization, and modeling of carbon nanotube-reinforced multiscale composites. *Composites Science and Technology* 69: 335–342.

Kitipornchai S, Chen D and Yang J. (2017) Free vibration and elastic buckling of functionally graded porous beams reinforced by graphene platelets. *Materials & Design* 116: 656–665.

Lei ZX, Liew KM and Yu JL. (2013a) Buckling analysis of functionally graded carbon nanotube-reinforced composite plates using the element-free kp-Ritz method. *Composite Structures* 98: 160–168.

Lei ZX, Liew KM and Yu JL. (2013b) Free vibration analysis of functionally graded carbon nanotube-reinforced composite plates using the element-free kp-Ritz method in thermal environment. *Composite Structures* 106: 128–138.

Lei ZX, Zhang LW, Liew KM, et al. (2014) Dynamic stability analysis of carbon nanotube-reinforced functionally graded cylindrical panels using the element-free kp-Ritz method. *Composite Structures* 113: 328–338.

Li W, Dichiara A and Bai J. (2013) Carbon nanotube–graphene nanoplatelet hybrids as high-performance multifunctional reinforcements in epoxy composites. *Composites Science and Technology* 74: 221–227.

Li X, Gao H, Scrivens WA, et al. (2004) Nanomechanical characterization of single-walled carbon nanotube reinforced epoxy composites. *Nanotechnology* 15: 1416–1423.

Liang J-Z, Du Q, Tsui GC-P, et al. (2016) Tensile properties of graphene nano-platelets reinforced polypropylene composites. *Composites Part B: Engineering* 95: 166–171.

Liew KM, Lei ZX, Yu JL, et al. (2014) Postbuckling of carbon nanotube-reinforced functionally graded cylindrical panels under axial compression using a meshless approach. *Computer Methods in Applied Mechanics and Engineering* 268: 1–17.

Lin F and Xiang Y. (2014) Vibration of carbon nanotube reinforced composite beams based on the first and third order beam theories. *Applied Mathematical Modelling* 38: 3741–3754.

Liu D, Kitipornchai S, Chen W, et al. (2018) Three-dimensional buckling and free vibration analyses of initially stressed functionally graded graphene reinforced composite cylindrical shell. *Composite Structures* 189: 560–569.

Malekzadeh P and Zarei AR. (2014) Free vibration of quadrilateral laminated plates with carbon nanotube reinforced composite layers. *Thin-Walled Structures* 82: 221–232.

Mallick PK. (2007) *Fiber-reinforced composites: materials, manufacturing, and design*: CRC Press.

Mareishi S, Rafiee M, He XQ, et al. (2014) Nonlinear free vibration, postbuckling and nonlinear static deflection of piezoelectric fiber-reinforced laminated composite beams. *Composites Part B: Engineering* 59: 123–132.

Matsunaga H. (2007) Vibration and buckling of cross-ply laminated composite circular cylindrical shells according to a global higher-order theory. *International Journal of Mechanical Sciences* 49: 1060–1075.

Mirzaei M and Kiani Y. (2016) Free vibration of functionally graded carbon nanotube reinforced composite cylindrical panels. *Composite Structures* 142: 45–56.

Mo CB, Cha SI, Kim KT, et al. (2005) Fabrication of carbon nanotube reinforced alumina matrix nanocomposite by sol–gel process. *Materials Science and Engineering: A* 395: 124–128.

Moradi A, Makvandi H and Bavarsad Salehpoor I. (2017) Multi objective optimization of the vibration analysis of composite natural gas pipelines in nonlinear thermal and humidity environment under non-uniform magnetic field. *Journal of Computational Applied Mechanics* 48: 53–64.

Naebe M, Wang J, Amini A, et al. (2014) Mechanical property and structure of covalent functionalised graphene/epoxy nanocomposites. *Scientific Reports* 4: 4375.

Nguyen N-D, Nguyen T-K, Nguyen T-N, et al. (2018) New Ritz-solution shape functions for analysis of thermo-mechanical buckling and vibration of laminated composite beams. *Composite Structures* 184: 452–460.

Nieto A, Lahiri D and Agarwal A. (2013) Graphene nanoplatelets reinforced tantalum carbide consolidated by spark plasma sintering. *Materials Science and Engineering: A* 582: 338–346.

Park M and Choi D-H. (2018) A two-variable first-order shear deformation theory considering in-plane rotation for bending, buckling and free vibration analyses of isotropic plates. *Applied Mathematical Modelling* 61: 49–71.

Patel BP, Ganapathi M and Makhecha DP. (2002) Hygrothermal effects on the structural behaviour of thick composite laminates using higher-order theory. *Composite Structures* 56: 25–34.

Pelletier JL and Vel SS. (2006) Multi-objective optimization of fiber reinforced composite laminates for strength, stiffness and minimal mass. *Computers & Structures* 84: 2065–2080.

Pérez-Bustamante R, Bolaños-Morales D, Bonilla-Martínez J, et al. (2014) Microstructural and hardness behavior of graphene-nanoplatelets/aluminum composites synthesized by mechanical alloying. *Journal of Alloys and Compounds* 615: S578–S582.

Phung-Van P, Abdel-Wahab M, Liew KM, et al. (2015) Isogeometric analysis of functionally graded carbon nanotube-reinforced composite plates using higher-order shear deformation theory. *Composite Structures* 123: 137–149.

Qaderi S, Ebrahimi F and Vinyas M. (2019) Dynamic analysis of multi-layered composite beams reinforced with graphene platelets resting on two-parameter viscoelastic foundation. *The European Physical Journal Plus* 134: 339.

Qiao P and Yang M. (2007) Impact analysis of fiber reinforced polymer honeycomb composite sandwich beams. *Composites Part B: Engineering* 38: 739–750.

Rafiee M, Nitzsche F and Labrosse MR. (2018) Modeling and mechanical analysis of multiscale fiber-reinforced graphene composites: Nonlinear bending, thermal post-buckling and large amplitude vibration. *International Journal of Non-Linear Mechanics* 103: 104–112.

Rafiee M, Nitzsche F, Laliberte J, et al. (2019) Thermal properties of doubly reinforced fiberglass/epoxy composites with graphene nanoplatelets, graphene oxide and reduced-graphene oxide. *Composites Part B: Engineering* 164: 1–9.

Rafiee M, Yang J and Kitipornchai S. (2013a) Large amplitude vibration of carbon nanotube reinforced functionally graded composite beams with piezoelectric layers. *Composite Structures* 96: 716–725.

Rafiee M, Yang J and Kitipornchai S. (2013b) Thermal bifurcation buckling of piezoelectric carbon nanotube reinforced composite beams. *Computers & Mathematics with Applications* 66: 1147–1160.

Rajoria H and Jalili N. (2005) Passive vibration damping enhancement using carbon nanotube-epoxy reinforced composites. *Composites Science and Technology* 65: 2079–2093.

Rashad M, Pan F and Asif M. (2016) Exploring mechanical behavior of Mg–6Zn alloy reinforced with graphene nanoplatelets. *Materials Science and Engineering: A* 649: 263–269.

Rashad M, Pan F, Hu H, et al. (2015) Enhanced tensile properties of magnesium composites reinforced with graphene nanoplatelets. *Materials Science and Engineering: A* 630: 36–44.

Reddy JN. (2004) *Mechanics of laminated composite plates and shells: theory and analysis*: CRC Press.

Rodrigues JD, Roque CMC, Ferreira, AJM, Carrera E and Cinefra M. (2011) Radial basis functions–finite differences collocation and a Unified Formulation for bending, vibration and buckling analysis of laminated plates, according to Murakami's zig-zag theory. *Composite Structures* 93(7): 1613–1620.

Roy T and Chakraborty D. (2009) Optimal vibration control of smart fiber reinforced composite shell structures using improved genetic algorithm. *Journal of Sound and Vibration* 319: 15–40.

Seidel GD and Lagoudas DC. (2006) Micromechanical analysis of the effective elastic properties of carbon nanotube reinforced composites. *Mechanics of Materials* 38: 884–907.

Sepahvand K. (2016) Spectral stochastic finite element vibration analysis of fiber-reinforced composites with random fiber orientation. *Composite Structures* 145: 119–128.

Shao LH, Luo RY, Bai SL, et al. (2009) Prediction of effective moduli of carbon nanotube–reinforced composites with waviness and debonding. *Composite Structures* 87: 274–281.

Shen H-S. (2004) Thermal postbuckling behavior of functionally graded cylindrical shells with temperature-dependent properties. *International Journal of Solids and Structures* 41: 1961–1974.

Shen H-S. (2012) Thermal buckling and postbuckling behavior of functionally graded carbon nanotube-reinforced composite cylindrical shells. *Composites Part B: Engineering* 43: 1030–1038.

Shen H-S, Lin F and Xiang Y. (2017a) Nonlinear vibration of functionally graded graphene-reinforced composite laminated beams resting on elastic foundations in thermal environments. *Nonlinear Dynamics* 90: 899–914.

Shen H-S, Xiang Y and Fan Y. (2018a) Postbuckling of functionally graded graphene-reinforced composite laminated cylindrical panels under axial compression in thermal environments. *International Journal of Mechanical Sciences* 135: 398–409.

Shen H-S, Xiang Y, Fan Y, et al. (2018b) Nonlinear vibration of functionally graded graphene-reinforced composite laminated cylindrical panels resting on elastic foundations in thermal environments. *Composites Part B: Engineering* 136: 177–186.

Shen H-S, Xiang Y, Lin F, et al. (2017b) Buckling and postbuckling of functionally graded graphene-reinforced composite laminated plates in thermal environments. *Composites Part B: Engineering* 119: 67–78.

Shen H-S and Zhang C-L. (2010) Thermal buckling and postbuckling behavior of functionally graded carbon nanotube-reinforced composite plates. *Materials & Design* 31: 3403–3411.

Shi D-L, Feng X-Q, Huang YY, et al. (2004) The effect of nanotube waviness and agglomeration on the elastic property of carbon nanotube-reinforced composites. *Journal of Engineering Materials and Technology* 126: 250–257.

Shin J-H and Hong S-H. (2014) Fabrication and properties of reduced graphene oxide reinforced yttria-stabilized zirconia composite ceramics. *Journal of the European Ceramic Society* 34: 1297–1302.

Shishesaz M, Kharazi M, Hosseini P, et al. (2017) Buckling behavior of composite plates with a pre-central circular delamination defect under in-plane uniaxial compression. *Journal of Computational Applied Mechanics* 48: 12.

Shojaee S, Valizadeh N, Izadpanah E, et al. (2012) Free vibration and buckling analysis of laminated composite plates using the NURBS-based isogeometric finite element method. *Composite Structures* 94: 1677–1693.

Shokrieh MM, Esmkhani M, Haghighatkhah AR, et al. (2014a) Flexural fatigue behavior of synthesized graphene/carbon-nanofiber/epoxy hybrid nanocomposites. *Materials & Design (1980-2015)* 62: 401–408.

Shokrieh MM, Ghoreishi SM, Esmkhani M, et al. (2014b) Effects of graphene nanoplatelets and graphene nanosheets on fracture toughness of epoxy nanocomposites. *Fatigue & Fracture of Engineering Materials & Structures* 37: 1116–1123.

Sobhani Aragh B, Nasrollah Barati AH and Hedayati H. (2012) Eshelby–Mori–Tanaka approach for vibrational behavior of continuously graded carbon nanotube-reinforced cylindrical panels. *Composites Part B: Engineering* 43: 1943–1954.

Song M, Kitipornchai S and Yang J. (2017a) Free and forced vibrations of functionally graded polymer composite plates reinforced with graphene nanoplatelets. *Composite Structures* 159: 579–588.

Song M, Yang J and Kitipornchai S. (2018) Bending and buckling analyses of functionally graded polymer composite plates reinforced with graphene nanoplatelets. *Composites Part B: Engineering* 134: 106–113.

Song M, Yang J, Kitipornchai S, et al. (2017b) Buckling and postbuckling of biaxially compressed functionally graded multilayer graphene nanoplatelet-reinforced polymer composite plates. *International Journal of Mechanical Sciences* 131-132: 345–355.

Su Y, Wei H, Gao R, et al. (2012) Exceptional negative thermal expansion and viscoelastic properties of graphene oxide paper. *Carbon* 50: 2804–2809.

Suhr J, Koratkar N, Keblinski P, et al. (2005) Viscoelasticity in carbon nanotube composites. *Nature Materials* 4: 134–137.

Tan H, Jiang LY, Huang Y, et al. (2007) The effect of van der Waals-based interface cohesive law on carbon nanotube-reinforced composite materials. *Composites Science and Technology* 67: 2941–2946.

Tehrani M, Boroujeni AY, Hartman TB, et al. (2013) Mechanical characterization and impact damage assessment of a woven carbon fiber reinforced carbon nanotube–epoxy composite. *Composites Science and Technology* 75: 42–48.

Thai CH, Ferreira AJM, Bordas SPA, et al. (2014) Isogeometric analysis of laminated composite and sandwich plates using a new inverse trigonometric shear deformation theory. *European Journal of Mechanics - A/Solids* 43: 89–108.

Thai CH, Ferreira AJM, Carrera E, et al. (2013) Isogeometric analysis of laminated composite and sandwich plates using a layerwise deformation theory. *Composite Structures* 104: 196–214.

Thai CH, Nguyen-Xuan H, Bordas SPA, et al. (2015) Isogeometric analysis of laminated composite plates using the higher-order shear deformation theory. *Mechanics of Advanced Materials and Structures* 22: 451–469.

Thai CH, Nguyen-Xuan H, Nguyen-Thanh N, et al. (2012) Static, free vibration, and buckling analysis of laminated composite Reissner–Mindlin plates using NURBS-based isogeometric approach. *International Journal for Numerical Methods in Engineering* 91: 571–603.

Thostenson ET and Chou T-W. (2002) Aligned multi-walled carbon nanotube-reinforced composites: processing and mechanical characterization. *Journal of Physics D: Applied Physics* 35: L77–L80.

Tserpes KI, Papanikos P, Labeas G, et al. (2008) Multi-scale modeling of tensile behavior of carbon nanotube-reinforced composites. *Theoretical and Applied Fracture Mechanics* 49: 51–60.

Wang A, Chen H, Hao Y, et al. (2018a) Vibration and bending behavior of functionally graded nanocomposite doubly-curved shallow shells reinforced by graphene nanoplatelets. *Results in Physics* 9: 550–559.

Wang F, Drzal LT, Qin Y, et al. (2015) Mechanical properties and thermal conductivity of graphene nanoplatelet/epoxy composites. *Journal of Materials Science* 50: 1082–1093.

Wang J, Liew KM, Tan MJ, et al. (2002) Analysis of rectangular laminated composite plates via FSDT meshless method. *International Journal of Mechanical Sciences* 44: 1275–1293.

Wang Y, Feng C, Zhao Z, et al. (2018b) Torsional buckling of graphene platelets (GPLs) reinforced functionally graded cylindrical shell with cutout. *Composite Structures* 197: 72–79.

Wang Y, Feng C, Zhao Z, et al. (2018c) Eigenvalue buckling of functionally graded cylindrical shells reinforced with graphene platelets (GPL). *Composite Structures* 202: 38–46.

Wang YQ, Ye C and Zu JW. (2019) Nonlinear vibration of metal foam cylindrical shells reinforced with graphene platelets. *Aerospace Science and Technology* 85: 359–370.

Wattanasakulpong N and Ungbhakorn V. (2013) Analytical solutions for bending, buckling and vibration responses of carbon nanotube-reinforced composite beams resting on elastic foundation. *Computational Materials Science* 71: 201–208.

Wernik JM and Meguid SA. (2014) On the mechanical characterization of carbon nanotube reinforced epoxy adhesives. *Materials & Design* 59: 19–32.

Wolff EG. (1979) Dimensional stability of structural composites for spacecraft applications. *Metal Progress* 115: 54–63.

Wu H and Drzal LT. (2014) Effect of graphene nanoplatelets on coefficient of thermal expansion of polyetherimide composite. *Materials Chemistry and Physics* 146: 26–36.

Wu H, Kitipornchai S and Yang J. (2015) Free vibration and buckling analysis of sandwich beams with functionally graded carbon nanotube-reinforced composite face sheets. *International Journal of Structural Stability and Dynamics* 15: 1540011.

Wu H, Kitipornchai S and Yang J. (2017a) Thermal buckling and postbuckling of functionally graded graphene nanocomposite plates. *Materials & Design* 132: 430–441.

Wu H, Yang J and Kitipornchai S. (2017b) Dynamic instability of functionally graded multilayer graphene nanocomposite beams in thermal environment. *Composite Structures* 162: 244–254.

Wu H, Yang J and Kitipornchai S. (2018) Parametric instability of thermo-mechanically loaded functionally graded graphene reinforced nanocomposite plates. *International Journal of Mechanical Sciences* 135: 431–440.

Wu HL, Yang J and Kitipornchai S. (2016) Nonlinear vibration of functionally graded carbon nanotube-reinforced composite beams with geometric imperfections. *Composites Part B: Engineering* 90: 86–96.

Xiao KQ, Zhang LC and Zarudi I. (2007) Mechanical and rheological properties of carbon nanotube-reinforced polyethylene composites. *Composites Science and Technology* 67: 177–182.

Yadav SK and Cho JW. (2013) Functionalized graphene nanoplatelets for enhanced mechanical and thermal properties of polyurethane nanocomposites. *Applied Surface Science* 266: 360–367.

Yang B, Kitipornchai S, Yang Y-F, et al. (2017a) 3D thermo-mechanical bending solution of functionally graded graphene reinforced circular and annular plates. *Applied Mathematical Modelling* 49: 69–86.

Yang B, Mei J, Chen D, et al. (2018) 3D thermo-mechanical solution of transversely isotropic and functionally graded graphene reinforced elliptical plates. *Composite Structures* 184: 1040–1048.

Yang B, Yang J and Kitipornchai S. (2017b) Thermoelastic analysis of functionally graded graphene reinforced rectangular plates based on 3D elasticity. *Meccanica* 52: 2275–2292.

Yang J, Ke L-L and Feng C. (2015) Dynamic buckling of thermo-electro-mechanically loaded FG-CNTRC beams. *International Journal of Structural Stability and Dynamics* 15: 1540017.

Yang J, Wu H and Kitipornchai S. (2017c) Buckling and postbuckling of functionally graded multilayer graphene platelet-reinforced composite beams. *Composite Structures* 161: 111–118.

Yang X, Wang Z, Xu M, et al. (2013) Dramatic mechanical and thermal increments of thermoplastic composites by multi-scale synergetic reinforcement: Carbon fiber and graphene nanoplatelet. *Materials & Design* 44: 74–80.

Yas MH and Samadi N. (2012) Free vibrations and buckling analysis of carbon nanotube-reinforced composite Timoshenko beams on elastic foundation. *International Journal of Pressure Vessels and Piping* 98: 119–128.

Yu T, Yin S, Bui Tinh Q, et al. (2016) NURBS-based isogeometric analysis of buckling and free vibration problems for laminated composites plates with complicated cutouts using a new simple FSDT theory and level set method. *Thin-Walled Structures* 101: 141–156.

Yue L, Pircheraghi G, Monemian SA, et al. (2014) Epoxy composites with carbon nanotubes and graphene nanoplatelets – Dispersion and synergy effects. *Carbon* 78: 268–278.

Zenkour AM. (2004) Buckling of fiber-reinforced viscoelastic composite plates using various plate theories. *Journal of Engineering Mathematics* 50: 75–93.

Zenkour AM and El-Shahrany H. (2019) Vibration suppression analysis for laminated composite beams embedded actuating magnetostrictive layers. *Journal of Computational Applied Mechanics* 50: 69–75.

Zenkour AM and Fares ME. (2001) Bending, buckling and free vibration of non-homogeneous composite laminated cylindrical shells using a refined first-order theory. *Composites Part B: Engineering* 32: 237–247.

Zhang LW, Lei ZX and Liew KM. (2015a) Free vibration analysis of functionally graded carbon nanotube-reinforced composite triangular plates using the FSDT and element-free IMLS-Ritz method. *Composite Structures* 120: 189–199.

Zhang LW, Lei ZX and Liew KM. (2015b) Vibration characteristic of moderately thick functionally graded carbon nanotube reinforced composite skew plates. *Composite Structures* 122: 172–183.

Zhang LW, Lei ZX, Liew KM, et al. (2014a) Large deflection geometrically nonlinear analysis of carbon nanotube-reinforced functionally graded cylindrical panels. *Computer Methods in Applied Mechanics and Engineering* 273: 1–18.

Zhang LW, Lei ZX, Liew KM, et al. (2014b) Static and dynamic of carbon nanotube reinforced functionally graded cylindrical panels. *Composite Structures* 111: 205–212.

Zhang LW, Zhu P and Liew KM. (2014c) Thermal buckling of functionally graded plates using a local Kriging meshless method. *Composite Structures* 108: 472–492.

Zhang Z, Li Y, Wu H, et al. (2018) Mechanical analysis of functionally graded graphene oxide-reinforced composite beams based on the first-order shear deformation theory. *Mechanics of Advanced Materials and Structures*: 1–9.

Zhao Z, Feng C, Wang Y, et al. (2017) Bending and vibration analysis of functionally graded trapezoidal nanocomposite plates reinforced with graphene nanoplatelets (GPLs). *Composite Structures* 180: 799–808.

Zhu P, Lei ZX and Liew KM. (2012) Static and free vibration analyses of carbon nanotube-reinforced composite plates using finite element method with first order shear deformation plate theory. *Composite Structures* 94: 1450–1460.

Zhu R, Pan E and Roy AK. (2007) Molecular dynamics study of the stress–strain behavior of carbon-nanotube reinforced Epon 862 composites. *Materials Science and Engineering: A* 447: 51–57.

Index

Printed in the United States
by Baker & Taylor Publisher Services

Printed in the United States
by Baker & Taylor Publisher Services